Global Perspectives on Migration and Development

Global Migration Issues

Volume 1

Series Editor:
Dr. Frank Laczko
Head of Research and Publications, International Organization for Migration (IOM), Geneva

This book series contributes to the global discussion about the future of migration policy through the publication of a series of books on emerging migration issues. Most reports on migration policy tend to focus on national or regional perspectives; books in this series will focus on global policy challenges, such as the impact of climate change or the global economic crisis, on migration.

This series is closely linked to the production of IOM's *World Migration Report*. Some of the books in this series will be based on research which has been prepared for the *World Migration Report*.

The series also includes a special focus on the linkages between migration and development, and the themes discussed each year at the Global Forum on Migration and Development (GFMD), given the growing policy interest in harnessing the benefits of migration for development.

For further volumes:
http://www.springer.com/series/8837

Irena Omelaniuk
Editor

Global Perspectives on Migration and Development

GFMD Puerto Vallarta and Beyond

Springer

Editor
Irena Omelaniuk
Senior Adviser
Global Forum on Migration and Development (GFMD)

INSTITUTO NACIONAL DE MIGRACIÓN

Editorial guidance by the Centro de Estudios Migratorios del Instituto Nacional de Migración de México

ISBN 978-94-007-4109-6 ISBN 978-94-007-4110-2 (eBook)
DOI 10.1007/978-94-007-4110-2
Springer Dordrecht Heidelberg New York London

Library of Congress Control Number: 2012937215

Printed on acid-free paper

Springer is part of Springer Science+Business Media (www.springer.com)

Foreword

This book draws inspiration from the international debate that has grown up around the Global Forum on Migration and Development (GFMD), specifically the outcomes of the fourth GFMD in Puerto Vallarta, Mexico, in 2010. The Forum is still young, having commenced in 2007 in Brussels, but it is growing and changing the way the world thinks about migration and development.

A joined-up debate about these two sensitive issues was long overdue and long in coming. However, within the past 5 years of the life of the Forum, we have made up quickly for decades of resistance and reserve about discussing migration in multilateral settings. The development context has in many ways made this more possible.

We are grateful to the former UN Secretary-General Kofi Annan for his vision and persistence in proposing this unique government-led global Forum in 2006 and to his successor Mr. Ban Ki-moon for maintaining the faith. After decades of inaction on this issue, the majority of governments at the UN High-Level Dialogue in 2006 agreed to commence global reflections on migration and development in such a forum outside, but linked to, the United Nations.

Today, the GFMD is the largest forum of its kind, engaging the interest and touching the policies, programmes and experiences of more than 160 countries from all regions and from all points on the migration and development spectrum. Each Forum meeting, ranging from Brussels to Manila, Athens, Puerto Vallarta and Geneva, has brought new issues to the table. Puerto Vallarta gave special attention to irregular migration, a hugely sensitive issue within societies and between countries, but one that so often results from inadequate or non-existent policies, or poorly implemented policies on the ground. Two critical foci of the Millennium Development Goals (MDGs) – family and gender – were also brought into the spotlight.

Importantly, Puerto Vallarta opened up a common space for governments and civil society to exchange views and ideas about contemporary challenges, such as public perceptions of migration and how these interconnect with public policy making. We learned how strategic partnerships between people, institutions, countries, regions, public and private sectors, and émigrés and their home communities can help narrow the gaps between the theory that migration and development are mutually reinforcing and the practical reality of that.

Many of these issues were taken forward by the Swiss chairmanship in 2011 in reflection of how the GFMD, through its dogged pursuit of outcomes and solutions around core issues, has evolved into a coherent process rather than just a string of disconnected annual conferences.

We are happy to present this book to the interested reader as a first coherent compilation of GFMD reflections on key issues at the interface of migration and development. We hope this book helps us understand better the overriding message of the GFMD, namely, that migration can bring untold benefits for human, social and economic development, when it is properly understood, managed and protective of people and families on the move.

Salvador Beltran del Rio
Commissioner
National Institute of Migration
(Instituto Nacional de Migración)
Mexico City

William Lacy Swing
Director General
International Organization for Migration (IOM)
Geneva

Acknowledgements

This volume is the first in a new book series launched by Springer Science + Business Media in partnership with the International Organization for Migration (IOM) entitled *Global Migration Issues.* The overall editor of the new series is Frank Laczko, Head of Research and Publications, IOM Geneva, whose guidance and comments have been critical to the preparation and completion of this book.

This book has also profited from the unflagging editorial guidance of Ernesto Rodriguez Chavez, Salvador Berumen Sandoval and Juan Carlos Calleros Alarcon, Centro de Estudios Migratorios (Center for Migration Studies, Mexico), and the vision and encouragement of Ambassador Juan Jose Gomez Camacho, Geneva.

The editors would like to thank especially all contributing authors for their commitment, expert input and forbearance, as well as Andrea Motta, Research Assistant at Georgetown University, without whose patient and expert copy-editing assistance this volume would not have made it to the publishers. Thanks also to Andrew Michaels and Antoinette Wills for their additional generous editing assistance.

The editor also wishes to thank the following persons for their substantive inputs and assistance in peer-reviewing the book (in addition to the authors for critiquing each other's chapters): Ricardo Cordero, Jean D'Cunha, Jerome Elie, Shahidul Haque, Frank Laczko, Blandine Mollard, Gloria Moreno-Fontes, Karoline Popp, Helen Schwenken, Ronald Skeldon, Federico Soda, Colleen Thouez, Elizabeth Warn and Rudolf Anich and Lisa Tsurufuji for assisting with the authors' workshop and GFMD symposium on the contents of the book's chapters.

Contents

Acronyms and Abbreviations

ADF	Anatolian Development Foundation
AFFORD	African Foundation for Development
AI	Amnesty International
ALFEA	Association of Licensed Employment Agencies
ARI	African Remittance Institute
AU	African Union
AVRR	Assisted Voluntary Return and Reintegration
BAIRA	Bangladeshi Association of International Recruiting Agencies
BMET	Bureau of Manpower Employment and Training
BMP	Building Migration Partnerships Project
BOESL	Bangladesh Overseas Employment and Services Limited
BPOs	Business Promotion Officers
BSR	Business for Social Responsibility
CAMPO	Cape Verde Centre for Migrant Support in the Origin Country
CARICOM	Caribbean Community
CEDAW	Committee on the Elimination of Discrimination Against Women
CIDA	Canadian International Development Agency
CMWF	International Convention on the Protection of the Rights of All Migrant Workers and Members of their Families
COD	Country of Destination
COO	Country of Origin
COT	Country of Transit
CSD	Civil Society Day
CSOs	Civil Society Organizations
DOFE	Department of Foreign Employment
EC	European Commission
ECOWAS	Economic Community of West African States
EoSB	End-of-Service Benefits
EPS	Employment Permit System
ES	Employment Services
EU	European Union

EU-LAC	EU-Latin American and Caribbean Dialogue
FAO	Food and Agriculture Organization of the United Nations
FDI	Foreign Direct Investment
FEA	Foreign Employment Act
FEPB	Foreign Employment Promotion Board
GARF	Portuguese Office for the Support of Family Reunification
GBA	Gender-Based Analysis
GCC	Gulf Cooperation Council
GCIM	Global Commission on International Migration
GDP	Gross Domestic Product
GFMD	Global Forum on Migration and Development
GMG	Global Migration Group
HIA	Health Impact Assessment
HLD	United Nations High-Level Dialogue
ICMC	International Catholic Migration Commission
ICMPD	International Centre for Migration Policy Development
ICMRW	International Convention on the Protection of the Rights of All Migrant Workers and Members of Their Families
IFAD	International Fund for Agricultural Development
IGC	Intergovernmental Consultations on Migration, Asylum and Refugees
IHH	Humanitarian Relief Foundation
ILC	International Labour Conference
ILO	International Labour Organization
INM	Mexico's National Immigration Office
IOM	International Organization for Migration
IPCC	Intergovernmental Panel on Climate Change
IRF	Inter-regional Fora
IT	Information Technology
JRC	Jesuit Refugee Service
LDCs	Least Developed Countries
M&E	Monitoring and Evaluation
M4MD	Markets for Migration and Development
MDGs	Millennium Development Goals
MENA	Middle East and North Africa
MERCOSUR	South America's Southern Common Market
MEWOE	Ministry of Expatriates' Welfare and Overseas Employment
MIDSA	Migration Dialogue for Southern Africa
MOU	Memorandum of Understanding
MRCs	Migrant Resource Centers
NAFEA	Nepal Association of Foreign Employment Agencies
NAPAs	National Adaptation Programs of Action
NEF	National Empowerment Foundation
NGOs	Non-governmental Organizations

OCW-Unifil	Union of Filipino Overseas Contract Workers
ODA	Official Development Assistance
OECD	Organisation for Economic Co-operation and Development
OECs	Overseas Employment Certificates
OFW	Overseas Filipino Workers
OHCHR	Office of the UN High Commissioner for Human Rights
OPAP	Overseas Placement Agencies of the Philippines
OPIs	Mexican Officers for the Protection of Infants
OWWA	Philippine Overseas Workers Welfare Administration
PERCO	Platform for European Red Cross Cooperation on Refugees, Asylum-Seekers and Migrants
PfPs	Platform for Partnerships
PICUM	Platform for International Cooperation on Undocumented Migrants
POE	Protectors of Emigrants
POEA	Philippine Overseas Employment Administration
PRSPs	Poverty Reduction Strategy Papers
QCEA	Quaker Council for European Affairs
RABIT	Rapid Border Intervention Teams
RCGM	Regional Consultations Group on Migration
RCM	Regional Conference on Migration
RCPs	Regional Consultative Processes
RECs	Regional Economic Communities
RELA	Corps Volunteers of the Malaysian People
RMCC	Regional Migration Coordination Committees
ROK	Republic of Korea
RSE	Recognized Seasonal Employers Scheme
SACM	South American Conference on Migration
SADC	Southern African Development Community
SAWP	Canada's Seasonal Agricultural Worker Program
SLBFE	Sri Lanka Foreign Employment Bureau
SMEDA	Small and Medium Enterprise Development Authority
SMEs	Small and Medium Enterprises
TPS	Temporary Protected Status
UAE	United Arab Emirates
UI	Unemployment Insurance
UK	United Kingdom
UN	United Nations
UNCED	UN Conference on Environment and Development
UNDP	United Nations Development Programme
UNEP	United Nations Environment Programme
UNFCCC	United Nations Framework for Climate Change Convention
UNFPA	United Nations Population Fund
UNGA	United Nations General Assembly
UNHCR	United Nations High Commissioner for Refugees

UNICEF	United Nations Children's Fund
UNIFEM	United Nations Development Fund for Women
UNODC	United Nations Office on Drugs and Crime
USA	United States of America
WEF	World Economic Forum

Chapter 1
Introduction: Making the Connections Between Migration and Development

Irena Omelaniuk*

Why Another Book on Migration and Development?

This book is one of the first to distill key research findings and unique features of the Global Forum on Migration and Development (GFMD) since its inception in 2007. Conceived in the United Nations (UN) General Assembly in 2006 as an informal, state-led, non-UN facility to discuss migration and development, the GFMD is the most prominent, wide-ranging dialogue process of its kind. Each annual meeting produces a report of proceedings, but by 2012 the Forum still had not published a comprehensive treatment of its major findings and outcomes. This is partly because the papers prepared for the Forum are usually tied to narrow, theme-specific roundtables.[1] But it may also be a reflection of the informal nature of the GFMD.[2]

---

* This chapter was prepared by Dr. Irena Omelaniuk, Senior adviser to the Global Forum on Migration and Development (GFMD), Switzerland, 2011.

[1] The inaugural GFMD meeting in Brussels in 2007 established a roundtable format for future government meetings, whereby governments form teams to jointly prepare the roundtables around themes discussed and agreed on an annual basis by the GFMD Steering Group (SG) and Friends of the Forum (FOF).

[2] The GFMD is not grounded in a constitution, but as an informal process is guided by a loose set of Operating Modalities agreed by governments in Brussels in 2007 (www.gfmd.org/en/process/operating-modalities.html). The Chair-in-Office each year is dependent on governments volunteering to host the Forum and usually alternates between developing and developed countries. The Chair-in-Office is supported by the voluntary supporting entities: the Troika, Steering Group, and Friends of the Forum. (The Troika comprises past, current, and future Chairs. The Steering Group consists of a mix of governments committed to offer sustained political, conceptual, and substantial support to the Forum process. The Friends of the Forum, open to all member states and observers of the United Nations, acts as a sounding board and advises on the agenda, structure, and format of each Forum meeting, including the thematic program for each year.)

I. Omelaniuk (✉)
Senior Adviser, Global Forum on Migration and Development (GFMD)
e-mail: Iomelaniuk@yahoo.com.au

I. Omelaniuk (ed.), *Global Perspectives on Migration and Development: GFMD Puerto Vallarta and Beyond*, Global Migration Issues 1, DOI 10.1007/978-94-007-4110-2_1,

The authors of this book, all directly involved in preparing background papers or participating in the GFMD discussions, are well placed to give fuller treatment to the themes discussed in the GFMD. They help set the specific GFMD issues in the global context and offer an assessment of the progress of the GFMD in advancing understanding of the linkages between migration and development.

The main point of reference of the book is the fourth GFMD in Puerto Vallarta 2010, which proposed some innovative policy thinking, subsequently pursued in some form by the Swiss GFMD 2011, in areas such as the social and financial protection of temporary migrants, gender and family, irregular migration, climate change, assessing policy impacts, government-civil society interactions, and governance of international migration in the development context.[3]

The book reflects the character of the GFMD by highlighting more the practical than the theoretical, and more action-oriented than discursive outcomes of the GFMD. It comments as much on the process of global debate on migration and development – between governments as well as among governments, civil society, and private sector actors – as on the issues themselves. It recognizes that how one organizes the debate on migration and development can greatly influence how to manage these two interlocking issues. In Puerto Vallarta and beyond, the Forum has tested some new forms of collaboration between government and civil society, ranging from Common Spaces in the plenary sessions of the summit government meetings to a symposium on "Joint Reflections on Migration and Development" among governments, civil society, and the authors of this book in Geneva, August 2011[4] (*graduateinstitute.ch/globalmigration/.../GFMD-joint-reflections.html*).

There is surprisingly little empirical evidence on where and how the connections between migration and development work at any point in the migration cycle, particularly in the more prevalent forms of temporary, circular, and mixed movements today. There is even less serious reflection on the global institutional architecture and the forums needed to close these gaps and push the debate forward in a concerted way toward global solutions (or at least solutions to global problems). The relative absence of development agencies, governmental and nongovernmental, in the global migration and development debate, also in the GFMD, may be part of the problem.

The book does not linger on the traditional development-related outcomes of migration per se, such as remittances, or diaspora and migrant investments. Instead, it focuses on a few strategic policy and institutional challenges, such as lower migration costs, social protections for mobile labor forces, gender and family-sensitive

---

[3] The Fourth Meeting of the Global Forum on Migration and Development, was hosted by Mexico in Puerto Vallarta, Jalisco, from November 8 to 11, 2010. It was attended by 131 countries and 38 international organizations. The overarching theme of the forum was "Partnerships for Migration and Human Development; Shared Prosperity, Shared Responsibility," which guided both the government and civil society meetings (http://www.gfmd.org/en/gfmd-meetings/mexico-2010.html).

[4] The symposium followed the pattern of brainstorming on past and possible future GFMD themes by governments, civil society, and the current organizers of the GFMD, as supported by MacArthur Foundation in 2009 and 2010.

programs, and basic protections for migrants in irregular circumstances. Alongside these issues, it considers the partnerships and frameworks of cooperation – regional consultative processes, bilateral and multilateral agreements, and common spaces between governments and civil society – that can make these configurations work.

Given the complexity of migration and development, there are no simple formulae and no intrinsically correct approaches, but a diversity of policies and programs that evolve through trial and error. Migration may well be the most rapidly changing global phenomenon today, and policies to manage migration need to be flexible and linked to related fields that increasingly determine and influence migration patterns. Governments are not managing migration alone; a vast array of players and factors – migrants, families, employers, labor recruiters, migration agents, labor demand, information technology, etc. – are driving and managing migration today.

The GFMD has discovered that the factors most influencing the positive or negative developmental effects of migration often have more to do with global labor markets, private sector practices, and information technology than migration policies. Managing migration has become more an issue of how to balance interests among the many stakeholders: employers' needs for skills and labor, individuals' needs for economic and social protection, and governments' needs to be seen to be controlling their borders and protecting national security. The study of migration today leads one to the study of economics, trade, labor, and social theories. Adding development to this equation has both complicated the debate and given an otherwise complex and sensitive issue a cohering framework.

The development facet has added more players to the field – development ministries, agencies, and practitioners – but also given a social and economic context to migration management arguments that most would not disagree with. The developmental benefits of migration bring governments, civil society, and private sector together; help combat negative public perceptions about migration; and even offer a rationale for migration as a proactive way of preparing for and responding to global crises.

This has compelled governments to work together more and seek joint approaches at the regional level through regional consultative processes (RCPs), but also at the global level, most notably through the GFMD. The significance of development in these discussions and the need for collaborative approaches have prompted one of the recurrent questions in this book: Where are the development agencies in the global migration and development debate?

Why Use the GFMD as a Reference Point?

The Process

There are few international forums that deal with the migration and development linkages from a practical angle. National and international development agencies mostly deal with mandated issues such as food security, sectoral development, etc., while migration agencies are mostly concerned with border integrity and migration management strategies. This is changing slowly, as knowledge of the more complex interplay between migration and development grows, inter alia through the work of the GFMD.

The GFMD is the largest and most multilateral and inclusive of all international processes dealing with these issues at both the substantive and institutional levels. Established by majority agreement of the UN member states, at the instigation of the then Secretary General, at the High-Level Dialogue on Migration and Development in New York in 2006,[5] it is the only global forum of its kind, both outside and within the UN system, for informal, nonbinding multilateral debate on this issue. The decision in 2006 to establish such a facility followed more than a decade of delayed action on Chapter X of the Programme of Action resulting from the 1994 International Conference on Population and Development.[6] The groundwork for both the establishment of an interagency global migration "facility" and the themes to be pursued by such a facility was laid by the Global Commission on International Migration (GCIM) set up in 2003 by the then UN Secretary General and a number of governments.[7]

While there had been little political appetite among many UN member states for a multilateral forum on migration in 1994, by 2006 governments were more prepared to test an informal, nonbinding, government-led variation on this. Since its establishment in 2007, governments have staunchly defended the informality and state ownership of the Forum. The GFMD brings together policy makers from developing and developed countries, and countries that host migrants and are the source of migrants. By the second year of its operation, Martin noted that the GFMD marked an "important transition in thinking about the governance of international migration from a purely unilateral to a multilateral framework for managing movements of people" (Martin 2008, p. 226). This transition may be best reflected in the increasing preoccupation of the Forum with relations between its government membership and civil society and international partners.

What makes the GFMD unique among international forums is its broad intergovernmental dynamic and practical orientation. The GFMD draws mainly on governments' ideas, experiments, and lessons learned on the ground, as well as evidence and ideas from nonstate expert organizations and civil society, to fill knowledge gaps about the interactions between migration and development and add value to policy formulation.

---

[5] The UN High-Level Dialogue on International Migration and Development, UN General Assembly, September14–15, 2006, was the first ever plenary session of the UN General Assembly on migration issues. The focus was on ways to maximize the development benefits of migration and to reduce difficulties. By agreement of the majority of Member States, this launched the Global Forum on Migration and Development as a state-led process. Belgium volunteered to host and organize the first meeting in 2007 (www.un.org/migration/).

[6] See Chapter X on International Migration, part A on International Migration and Development, in the ICPD 1994 Programme of Action: http://www.ippf.org/en/About/ICPD±Programme±of±Action.htm.

[7] The GCIM comprised 19 commissioners from all walks of life, mandated to formulate a coherent, comprehensive, and global response to the issue of international migration (www.gcim.org/en/).

While the issues it discusses may not be new, the way it draws more than 160 governments and their nonstate partners together to discuss those issues is unique. The greatest value of the Forum, therefore, may well lie in the process, which is informal and inclusive, and its emphasis on learning about good practices from practice rather than through doctrine or assumptions.

The Issues

The Forum has worked with a broad concept of development, directed in earlier years more at economic development, brain drain, remittances, contributions of diaspora, etc., lowering recruitment costs, low-cost migrant loans, etc., but shifting more recently to human development. This was in part evolutionary, but partly also reinforced by the 2009 UNDP Human Development Report on *Overcoming barriers: Human mobility and development.* A more multidimensional concept of development, including human, economic, and social growth, opened the way for the Greek GFMD Chair in Athens, 2009, to discuss one of the most sensitive migration issues for many destination countries – migrant integration:

> The UNDP Human Development Report's concept of human development as a process of enlarging people's choices… by expanding human capabilities and functionings, may be a useful starting point when examining the relevance of integration policies for development. Also the UNGA Agenda for Development defines development as a multidimensional undertaking to achieve a higher quality of life for all people (GFMD 2010).[8]

It is interesting to observe the fortunes of a process like the GFMD as the UN system reviews its approach to the Millennium Development Goals (MDGs), the guiding framework for international development (www.un.org/millenniumgoals/). The MDGs have not featured prominently in the GFMD, although most of the goals are indirectly addressed in some form (poverty reduction, gender, education, reduced child mortality, maternal health, environment, global partnerships).

The GFMD reflections on development in the migration context have been more general and broad-ranging than the MDGs, extending to crisis scenarios such as climate change. Within the general global MDG targets, the Forum has identified specific examples, also at local and regional levels, where migrant-oriented policies could improve personal and family well-being, as well as national and global economies. For example, social welfare benefits for globally mobile workers through contributory schemes that obligate the migrant and the employer could offer more lasting benefits to the migrant and family after the worker returns home, especially where there is agreement on portability of the benefits. At the front end, more rational admissions policies and matching of skills with jobs can enhance

[8] See the Background Paper for Roundtable Session 2.1 on "Inclusion, protection and acceptance of migrants in society – linking human rights and migrant empowerment for development," GFMD, 2009, Athens.

the outcomes for overseas contract workers and protect them from possible irregular migration situations. These are sensible strategies for both the social and economic effects of migration.

The fourth GFMD in Puerto Vallarta, Mexico, in 2010 set some fresh markers for the migration and development debate. It opened the Forum to issues that had been peripheral to the process, yet are key to understanding the interactions between migration and development. These include irregular migration, gender, family, and the often ambiguous role of the private sector in facilitating and managing labor mobility between high- and lower-income countries.[9]

In Puerto Vallarta, the outcomes of the Forum begin to crystallize more around agreed sets of principles or "architectures," for example, to protect and empower temporary and circular labor migrants. These include actions at strategic points in the migration life cycle from predeparture to return that the GFMD has continued fleshing out in 2011: predeparture information, regulated recruitment, skills-job matching, low-cost loans, contract validation, social security, affordable health care, one-stop shop processing, skills recognition, gender sensitivity, family support, and incentives for return and reintegration (see roundtable 2 on "Human Mobility and Human Development").

This book examines some of the most concrete preoccupations and outcomes of GFMD 2010 and beyond and reflects on how these may be changing the way the world thinks and acts on migration and development. It draws from government sources, current research, the roundtable debates, the Summary of Proceedings of the Mexico GFMD 2010, and related follow-up during the Swiss GFMD in 2011.

The book demonstrates how bringing the public and private worlds to the same table around some common challenges may make the process of the GFMD as important, perhaps more so, as any actual policy reforms. It points to real impacts the Forum has had on the agendas of other related international and regional forums and to evidence of its cohering effects on the international debate on migration and development (GFMD 2010 Summary of Proceedings).

But the book also asks why development agencies or diaspora organizations, both key to the migration and development connection, are still largely absent from the process and from immigration policy making in most countries. Research has shown that often the countries of interest for development agencies do not coincide with those of interest to migration agencies, nor can budgets tied to core development mandates be so easily shifted to meet migration concerns, and, finally, development agencies are concerned not to dilute or place conditions on development aid by linking too closely to migration management imperatives. The book also speculates whether the Forum is nevertheless strengthening multilateral cooperation toward more coherent global governance of migration.

---

[9] The overarching theme of the Puerto Vallarta Forum was *Partnerships for Migration and Human Development: Shared Prosperity – Shared Responsibility.* In support of this theme, three roundtables were conducted on (1) the flagship theme (covering regular and irregular forms of migration), (2) Human Mobility and Human Development (covering the protection of labor migrants and gender issues), and (3) Policy and Institutional Coherence to Address the Relationship Between Migration and Development (http://www.gfmd.org/en/documents-library/mexico-2010.html).

The following tour of the chapters draws out the major issues at the heart of the Forum today and of the global debate on migration and development generally. The chapters demonstrate some incremental advances in awareness about the interconnected nature of migration and development and good practices in legislation, policy, and operations that strengthen the positive links, for example, in temporary/circular forms of labor migration, or through Poverty Reduction Strategy Papers (PRSPs) that include migration, or through a new generation of Migration Profiles that support development planning.

Reducing Recruitment Costs Can Generate More Benefits for Low-Skilled Migrants

The GFMD recognized from the outset that high front-end costs of migration, particularly for labor recruitment, can negatively impact the well-being, productivity, and potential of migrants to lift their families out of poverty. These costs range from visa, passport, and medical fees to travel and labor recruitment fees. In some parts of the world, these costs can account for a third of the expected earnings of overseas contract workers in their first 3 years abroad. Such high expenses for low-income earners can stifle mobility and cause migrants to become so indebted that they violate the terms of their visas by taking on second jobs or overstaying, or send fewer remittances home to achieve savings targets, and contribute less to development.

In Puerto Vallarta in 2010, the GFMD discussed how to reduce the costs of migration, based on a study by Philip Martin commissioned by the GFMD Working Group on Protecting and Empowering Migrants for Development.[10] The goal was to test the feasibility of NGO-bank partnerships to provide low-cost loans to lower-skilled Bangladeshi workers headed for the Gulf Cooperation Council (GCC) States.[11]

The GFMD has consistently sought practical ways to lower migration costs, particularly the high costs of labor recruitment. The focus on recruitment fees relates to the fact that, among the many costs of migration, these were amenable to government intervention, particularly for low-skilled workers. Martin notes that reducing recruitment costs can generate more benefits for low-skilled migrants than further lowering the costs of remittances, while generating significant development-related gains for migrants, employers, and governments. However, lowering recruitment costs may be harder than reducing remittance costs, because job matching is more complex than sending money over borders.

---

[10] In 2009, GFMD Member States agreed to the establishment of two ad hoc working groups to assist future Chairs-in-Office in prioritizing and taking forward the outcomes from the annual GFMD meetings: (1) on Protecting and Empowering Migrants for Development (cochairs Philippines and United Arab Emirates) and (2) on Policy Coherence, Data and Research (cochairs Morocco and Switzerland): www.gfmd.org/en/adhoc-wg/protecting-and-empowering.html

[11] The study, commissioned by the UAE, cochair of the Working Group on Protecting and Empowering Migrants for Development, examines labor recruitment in Asia (Martin 2009).

The challenge facing policy makers is to design systems that protect migrants – particularly low-skilled workers – from exploitative intermediaries, while avoiding overregulation of an industry that can play a useful role in protecting migrants. Martin targets four areas for fresh policy action: matching the worker to the job, dealing with the layers of local intermediaries operating under the radar screen, costs of recruitment, and regulations and realities.

Looking at Asia (Bangladesh, India, Nepal, Philippines, and Sri Lanka), this chapter shows that most countries sending workers abroad already have laws and regulations that require recruiters to register, obtain licenses, and not exceed maximum fees charged to workers. But these regulatory provisions mostly do not extend to local subagents, who may charge extra or unduly high fees. This is ascribable to a number of factors, including corruption in what has become a multibillion-dollar industry, inadequate information and training of migrants, legal difficulties with enforcement across borders, and the absence of longer-term, enduring relationships between recruiters, migrants, and foreign employers (which can encourage fraudulence and predeparture debts).

One solution to the challenges posed by the private recruitment industry are government-to-government arrangements that exclude intermediaries or involve nonprofit partners such as IOM or NGOs. Codes of conduct for recruiters are another option, and host countries and employers can consider mechanisms to give preferential access to migrants who arrive through recognized and licensed recruiters. But codes of conduct are difficult to enforce across borders, and overregulation often has the opposite effects intended, namely, to drive migrants and/or agents underground and discourage private sector entrepreneurship.

Martin embeds recruitment solutions in a package of policies to reduce or eliminate predeparture recruitment debts; enable migrants to increase their savings and remittances; reduce the need for irregular migration; satisfy the employer, the governments, and the economy; and better prepare the workers to return home. Government actions include educating and informing the migrants, regulating recruiters, promoting competition and/or ethical recruitment among recruiters, or establishing government recruiting monopolies. Development that offers decent work at home is the ultimate answer to corruption in the recruitment process. However, smart admissions programs at destination that match workers to real jobs may also undercut the profitability of bad recruiters.

Beyond Puerto Vallarta, some advances have been made in Asia toward lowering the costs and risks to overseas contract workers in the Gulf States. In 2011, two follow-up workshops were held in Dubai and Dhaka on these issues, and agreement was reached among governments on a set of principles and good practices in the areas of prerecruitment, obtaining and protecting a contract, and regulating recruiters (with incentives). The Dubai workshop also yielded a draft framework of regional cooperation on recruitment practices for consideration in the context of Abu Dhabi Dialogue.[12] The UAE Ministry of Labor has implemented new laws and practices to

[12] See a description of the dialogue in the Chap. 13 by Achieng on "Regional and Inter-Regional Processes: Advancing the Discourse and Action on Migration and Development."

prohibit UAE employers from charging recruitment fees, to protect overseas workers' wages by requiring employers to deposit wages in bank accounts, and to protect workers from contract substitution or alteration through electronic transfer of the contracts from country of origin to country of destination.[13]

Circular Migration: A GFMD Mantra or Just a Good Practice?

The GFMD has had a special preoccupation with circular migration as a vehicle for increasing the development gains of labor mobility for countries of origin and destination, and for migrants and their families.[14] The concept was that people who circulate stay in touch with the countries and communities they come from and stimulate beneficial returns and exchanges between countries. Circulation can have short-term benefits and longer-term drawbacks for the migrants and families. Some governments from the outset saw the potential of circular labor migration to contribute to their national development efforts; others felt that circular migration was being confused with limited, exploitative forms of temporary labor migration.

One of the enduring legacies of the Forum may well be the fact that governments have agreed on a broad, flexible definition of "circular migration," which recognizes the benefits of facilitating spontaneous mobility between countries in temporary, long-term, or permanent ways.[15] Over time, the GFMD focus has shifted from trying to understand the difference between temporary and circular migration to the social and financial security measures that protect overseas workers in any kind of temporary or circulatory situation.

In Brussels in 2007, Mauritius offered to the GFMD a circular migration model embodied in bilateral agreements negotiated with France and Canada. The value of this model for the GFMD is that it (a) exemplifies how circular migration can be an integral part of national development planning and (b) shows how circular migration in a globalized labor market can bring origin and destination countries together as equal business partners around mutual benefits. This circular migration model, which involves training at point of origin for employment in the receiving country, is featured in the toolkit of good practices in the Forum's web-based Platform for Partnership (see Textbox 1).

---

[13] See the Summary Report on the Workshop on Recruitment of Workers for Overseas Employment, Dubai, January 18–19, 2011, and the Summary Report on the Workshop on Lowering Migration Costs in Collaboration with, and Developing a Model Bilateral Framework with Receiving Countries, Dhaka, September 21–22, 2011, www.gfmd.org.

[14] See the definition agreed by participating governments in the first GFMD roundtable debate on this in 2007: "Circular Migration is the fluid movement of people between countries, including temporary or more permanent movement which, when it occurs voluntarily and is linked to the labor needs of countries of origin and destination, can be beneficial to all involved."

[15] At the GFMD meeting in Brussels in 2007 and the follow-up workshop in Mauritius in 2008, governments recognized that permanent residence, citizenship, and dual and multiple citizenship can spur more spontaneous circulation.

Many GFMD member states, including Costa Rica, the Netherlands, Ukraine, Portugal, and Sweden, have changed their legislation or introduced programs to test circular migration programs. Some were part of the European Union's efforts to tie labor market needs to development needs in countries of origin through mobility partnerships.[16] In Sweden, an Independent Parliamentary Committee on Circular Migration and Development was set up in 2009 to examine the relationship between circular migration and development. The Parliamentary Committee supports the concept of "spontaneous circular migration," that is, voluntary, regular circulation facilitated by, for example, multiple reentry visas or dual or multiple citizenship, and recommends a range of flexible policies and support services to achieve this (see Textbox 2).

The GFMD Working Group on Protecting and Empowering Migrants for Development in 2010 examined ways to minimize the social and financial risks for temporary/circular migrants. One way was through the comprehensive policy package outlined by Martin above, but another was through smarter social protection measures. The discussion moved from a definitional dispute to the more practical issue of how to better protect and support those who are already circulating their skills and services globally.

How Can Social Protection Serve Temporary Migrants Best?

Social protection, pensions, and other social benefits and assistance offered by governments are part of the cost equation in any migration cycle. If lower recruitment costs help raise savings and remittance potential, and lower the risk of irregular migration, public welfare support can lower the social risks and costs for lower-income families. Social protection is an important plank in any comprehensive policy to protect and empower migrant workers and their families, and to leverage the developmental outcomes of their migration.

Social protection for mobile workers, particularly low skilled, has received relatively little attention from policy makers and studies on migration and development. The Chap. 5 by Holzmann and Pouget reflects on the discussions in Puerto Vallarta and a background study on this commissioned by the GFMD Working Group on Protecting and Empowering Migrants for Development (Holzmann and Pouget 2010).

Work-related benefits and services, ranging from preemployment training, housing, and transport to unemployment benefits, vary widely across countries of destination and origin and are often absent entirely. Family benefits, such as health care or family reunification, are mostly not available to temporary migrant workers in countries of

[16] See Commission of the European Communities (2007).

destination. This has caused some countries of origin to offer their own low-cost welfare package from home. Pension benefits are available in OECD countries, but most others provide no pension benefits, or offer end-of-service benefits, which as Holzmann and Pouget report have a mixed record of success.

Most development economists, social protection experts, and international conventions call for temporary migrants to have the same access to social programs as local workers. But this often requires special bilateral or multilateral agreements, such as for portability of benefits. Migrants may require additional services such as predeparture training, accommodation assistance, and financial services, and in certain cases may not qualify for benefits such as unemployment benefits and pensions, because of a qualifying period beyond their contract (e.g., with seasonal workers). Social programs for temporary migrants thus need to be adjusted to their idiosyncratic risk profile and needs.

The challenge is how to make those adjustments in ways that protect and support the migrants and are cost-effective for all involved, particularly where there are few or no programs already available in either the country of origin or destination.

The Chap. 5 by Holzmann and Pouget closes the gap on social protection of temporary workers crossing borders by taking stock of existing social benefits for temporary or circulating cross-border workers and offering a framework concept against which to assess existing social protection schemes. The study informing the chapter (Holzmann and Pouget 2010) covered new territory, both for the GFMD and the global debate on migration and development, namely, social protections in an age of rapid and complex global labor mobility. It endeavors to apply these to low-skilled migrant workers.

The authors are realistic about the likelihood of seasonal workers receiving unemployment or pension benefits, for which they are likely to be ineligible due to a mandatory qualifying period. For unemployment and pension benefits, they suggest portability via bi- and multilateral agreements and changes in benefit design and/or a move toward individual accounts. The authors propose as an alternative to end-of-service benefits funded contribution (pension) schemes for all temporary migrant workers. More broadly, this chapter argues that savings options for migrants can be improved through appropriate financial instruments for retirement savings, strict regulation and supervision of financial institutions, safeguards for investors/migrants, and a minimum financial capability for migrant investors to make basic choices.

This chapter offers some possible next steps to deepen the policy analysis on social protection for temporary migrants, including extending the inventory by Holzmann and Pouget to medium-intensity migration-receiving countries and undertaking a full review of legal provisions in countries of origin. They urge that policy makers move from simply reviewing legal protection to evaluating actual protection, for example, through rigorous monitoring and evaluation to yield the empirical data necessary to tailor social protection to the specific needs of different temporary, circulating global workers.

Health Care: The Missing Element in Protecting Temporary Labor Migrants?

The GFMD has highlighted the fact that health care for the many types of migrants today remains elusive, despite a growing awareness of the general benefits to society of granting migrants access to existing public health care services: cost-effectiveness, human rights protection, social integration, public health, and commitment to international resolutions and declarations.[17] Countries like Mexico and Sri Lanka are ensuring better access to health care services for their own nationals while abroad, while many other countries hosting migrants – including middle-income countries such as Thailand and Argentina – have developed schemes to give nonnationals in their countries better access to health care services. The Sri Lankan Government is developing a national policy on migration health and has established an extensive interministerial and interagency coordination framework for migration health development (see Textbox 3).

Gender and Family: A New Way of Looking at Migration and Development?

Despite growing evidence about the importance of gender and family for development, there is little international discussion of the concrete interfaces between these three fields, and even less so in the migration context. After minimal attention to these issues by earlier GFMD meetings, the Puerto Vallarta forum placed gender and family squarely on the agenda as a way of looking differently at the critical links between migration and development.

Calleros' chapter shows how gender and family are pivotal to achieving the Millennium Development Goals, and applying gender and family as units of analysis is in line with more recent UN efforts to redefine development from the more human/social angle. Migration studies and policies traditionally have focused on individuals rather than on the family as a unit of analysis. Without replacing this focus, participants in the Puerto Vallarta meeting saw gender and family as a way of adding new perspectives to the analysis.

The most vulnerable groups during migration are likely to be children and aged persons. Immigration processes increasingly involve children, both in regular and irregular migration situations. Yet there is a serious dearth of information about these family members to inform policies and a lack of policy consideration of these groups. The plight of children and the elderly brings to light the fact that poor or nonexistent social policies in countries of origin are one of the biggest causes of

[17] Including 61st World Health Assembly Resolution (WHA 61.17, 2008), which calls upon Member States to promote the "health of migrants."

irregular migration, including trafficking in persons which can undercut the positive effects of migration for development. This is clearly a broader development issue with migration consequences.

Calleros' chapter reflects the efforts in Puerto Vallarta to identify where such policies to protect and support the most vulnerable at home and abroad were being applied. How were concerns about family unity and gender sharpening understanding of the development-related causes and consequences of migration? How were they being factored into coherent policy making? And how to test whether such special policy considerations can amplify the developmental effects of migration?

Using the "transnational family" as a unit of analysis of the effects of migration policy can help us to recognize the transformations and fragmentation that family structures undergo as a result of complex global labor dynamics today. This in turn helps uncover the inadequacies of immigration and residency policies that, for example, give a child residence status (for being born there) and a parent deportation orders for being undocumented, within the same family. Mixed-status families in the same location are a new migration reality that challenges the traditional concept of family and the policies tailored to this concept.

New policies and practices are needed in both countries of origin and destination to preserve the principle of family unity, a fundamental right under international law. Some good practices have emerged at the country of origin end, such as comprehensive family counseling, welfare services, skills upgrading, advocacy abroad, family circles back home, and general psychosocial preparation for return and reintegration. Partnerships with IT companies can enable émigrés to maintain their family ties through Internet applications.

Still too few countries consider the needs of families, spouses, and children who stay behind, or accompany the principal migrant. Some stringent immigration policies even cause hardship and increase the vulnerability of family members; others limit access to labor markets for women. Intended or unintended gender-biased and gender-blind immigration policies can reinforce preconceived gender roles and power distribution within traditional cultures. Empowering women migrants can be a critical factor for family growth, community development, and poverty reduction. But this potential can be curtailed if policy makers in countries of origin and destination are not informed about the value of gender-testing their policies.

Calleros' chapter opens the way for further exploration of collaborative transnational actions between civil society and governments to support and protect women, families, and particularly children, and of policies to strengthen such action in the future. As a good practice, for example, Mexico has established a specialized Corps of Child Protection Officers and developed a training module for migration officials dealing with child rights protection, that it has offered to migration officers from other countries. This program is now showcased on the GFMD Platform for Partnerships both as a good practice and to attract willing partners in training from other countries (www.gfmd.org/pfp).

The Special Case of Global Domestic Workers

GFMD 2010 opened a debate on the special case of global domestic workers to highlight the development challenges of labor mobility, including the gender and family challenges.[18] Following the recommendations of Puerto Vallarta, the Swiss GFMD in 2011 embarked on a series of global thematic meetings, in Jamaica and Ghana, to discuss how governments, domestic worker associations, other civil society groups, and international organizations could take joint actions to better protect and support domestic workers and their families. These were timely, in view of the adoption of ILO Convention No 189 on Decent Work for Domestic Workers at the International Labour Conference (ILC) in June 2011 (see Textbox 4).[19]

Irregular Migration: Causes, Patterns, and Strategies

Irregular migration, a major preoccupation of the Puerto Vallarta GFMD,[20] is both an issue of costs and of rights. Castles et al. observe that irregular migration status per se does not necessarily prevent migrants and their families from benefiting economically from their mobility, but irregular migrants may not achieve as much human development. When undocumented migrants are marginalized, excluded from support services, decent wages and work conditions, banking and remitting facilities, etc., their capacity to lift their families out of poverty or socioeconomic deprivation is severely curtailed.

While only 10–15% of migrant populations are estimated to be in some kind of "irregular" status,[21] the negative impacts of irregular status on individual migrants' well-being can be high. The exact connections between irregular migration and development, and the negative effects of irregular status on development, are variable, but they include loss of income, social support, skills and labor potential, and remitting powers. The Puerto Vallarta discussions were based on the assumption that legal, safe, and empowering migration is the best way to prevent irregular migration also remains largely untested.

---

[18] See the Annex to the Roundtable 2.2 background paper on "Uncovering the Interfaces Between Gender, Family, Migration and Development: The Global Care Economy and Chains," www.gfmd.org

[19] To be considered in conjunction with the General Comment on Migrant Domestic Workers by the Committee on the Protection of the Rights of All Migrant Workers and Members of Their Families (CMW) 2010 and the Committee on the Elimination of Discrimination Against Women (CEDAW) General Recommendation No. 26 on Women Migrant Workers 2008.

[20] See Roundtable 1.2 on "Joint Strategies to Address Irregular Migration," GFMD Puerto Vallarta, November 8–11, 2010.

[21] See the definition of "irregular migration" by Castles et al. in the Chap. 9 "Irregular Migration: Causes, Patterns and Strategies."

Castles et al. reinforce this by arguing that many of the problems arise through lack of adequate migration policies, not with the migrants themselves. Migrants in irregular circumstances need basic protections, and the conditions leading to their situation should be seriously addressed. To take the example of labor mobility, if states do not create legal migration systems despite strong employer demand for workers unavailable locally, migrants opt for irregular solutions and experience high levels of risk and exploitation, including at the hands of smugglers and traffickers. Only the combined actions of government, civil society, private sector, and migrants can best address the causes and outcomes of irregular migration, while protecting the migrants and their families.

GFMD has touched on irregular migration without controversy or rancor, in part because of the practical, solutions-oriented approach taken by governments. Governments recognize that they have a shared interest in addressing the root causes (demography, labor demands, economic inequality, development needs, etc.), for example, through better management of demand and supply of human resources. The "migration life cycle" perspective adopted by the GFMD has also been more helpful in this regard than the usual dichotomy of country of origin vs. country of destination. It has also helped raise understanding of the "mixed flows" phenomenon.

But policies are still too much influenced by perceptions and politics, rather than facts and data. There are still too many distortions regarding the impacts of irregular migration on host societies (e.g., nonpayment of taxes, drains on employment, wage reduction, abuse of social security systems, etc.). There are also still few policy makers, including in the GFMD, who have confronted the complex issue of relative costs, for example, of managing irregular migration vs. not managing it, or of full rights-based migration programs vs. lower-scaled rights-based forms of labor mobility, of the costs to migrants working abroad in irregular situations vs. staying at home, or of stay-at-home development investments vs. labor export.

Castles et al. question the traditional emphasis on surveillance, law enforcement, detention, and capacity building in policy packages to combat irregular migration. Instead, they emphasize practical measures, including more regular migration avenues such as temporary or circular migration programs, respect for human rights, civil society actions, and prosecution of criminal organizations involved in trafficking and smuggling. States' involvement in the protection of the human rights of migrant workers in circular and temporary migration schemes is crucial. Where this is not happening, trade unions, migrant associations, and NGOs play a major role in supporting, protecting, and advocating for the migrants (Avendaño 2009, p. 13–16).

This chapter proposes, as an important first step, to understand the reality of irregular migration and counter stereotypes that link migration to crime, insecurity, and terrorism. One way forward may well be to highlight the contribution of irregular migrants to both countries of origin and destination, as discussed in Puerto Vallarta. The chapter notes that, at a time when developed countries and newly industrializing countries are competing for scarce skills, the labor-importing countries have begun to talk to governments of migrant-origin countries through such mechanisms as the GFMD.

Climate Change and Its Impacts on Migration and Development

Climate change is relatively new both to the GFMD and the international discourse on migration and development generally. Yet it represents an increasingly imminent interface between the two policy fields. It was tabled at the Puerto Vallarta GFMD in 2010 as an emerging issue that seriously challenged policies, institutional coherence, and cooperation across many sectors. The connections and implications for development can be tenuous.

While the impacts of climate change on migration have been debated for almost 20 years, there continues to be an insufficiently informed debate on the links with development, both as cause and effect. The Puerto Vallarta GFMD was an exception to this, as climate change was discussed as a factor among many that drive current migration. Martin and Warner observe that migration as a response to climate change is difficult to separate from other so-called drivers or factors of population movement. Any population that may be impacted by changing climate is already being affected by forces that fuel migration. Environmental factors are seldom the most important causes of migration, but in combination with poverty, insecurity, and poor governance, climate change may be a trigger for large-scale movements.

Martin and Warner review current knowledge about how climate change may affect migration, and the implications for adaptation, development, and immigration policies. Drawing on discussions at the GFMD and the UN Framework Convention for Climate Change summit in Cancun, the authors examine how climate change can affect migration (and development) through four specific channels: changes in weather patterns, rising sea levels, increased frequency and magnitude of weather-related natural hazards, and competition over natural resources that can lead to conflict and precipitate movements. Good practices are considered that include migration and development in longer-term research and prognostications about climate change.[22]

Most governments do not link their migration policies to climate (or environmental) change concerns. Similarly, most governments do not include migration in their National Adaptation Programmes of Action (NAPAs) and other development planning tools, at least not as a proactive strategy to plan for and adapt to environmental change. Traditional approaches to national disasters are still more oriented to humanitarian responses than to forward and proactive planning. While expert agencies like the World Bank have written about the risk-aversion and shock-absorbing effects of migrant remittances and other migrant and diaspora investments in home communities, these findings are not in national development or national preparedness strategies. This would clearly require a more joined-up approach

[22] Martin and Warner point to the UK Government's "Foresight" project on Global Environmental Migration as demonstrating some important findings and lessons in coherence. The project explores the links between climate change and migration and the challenges over the coming decades using scenario planning and other scientific tools. The purpose is to inform and assist decision makers to understand the future impacts of their decisions today (http://www.bis.gov.uk/foresight/our-work/projects/current-projects/global-migration).

within government and between government and other affected partners across migration, developmental, and humanitarian actors.

An interdisciplinary approach is even more critical for data collection, research, modeling, and capacity building. Since these are also mantras for general good governance on migration and development, a key message here may well be to ensure sound policies and systems for forward planning and responses are in place permanently. As in most areas explored by the GFMD, high-quality data on migration and climate change, and the ability and resources to collect and analyze them, are seriously lacking in most countries.

Martin and Warner recommend investments in dialogue and exchange of best practices; participatory policy planning; and proactive approaches to adaptation strategies, effective disaster risk reduction, conflict mediation, and disaster management programs to reduce the likelihood of emergency movements with accompanying humanitarian consequences. The chapter points the way to improving the quality of data and research on climate change, migration, and development and setting future priorities. It increases awareness of what can be done to manage risks in vulnerable zones, communicate information to populations who may be at risk, and contingency plan for possible relocation or resettlement. If not addressed, this phenomenon could lead to riskier forms of irregular migration.

The GFMD may not be a primary forum to deal with climate change, but it has brought a new perspective and proactive role for migration into the global debate on climate change. Climate change is one area where the GFMD is having an impact on the deliberations of other major global forums. As Martin observes, despite the nonbinding nature of GFMD conclusions, there has been some follow-up action on recommendations out of Puerto Vallarta in 2010 at the June 2011 Nansen Conference on Climate Change and Displacement in the 21st Century in Oslo.

Impact Assessments and Migration Profiles: Closing the Gaps in Coherence?

Every GFMD meeting since 2007 has called for more robust data and analysis of the linkages between migration and development. This has been part of a larger GFMD concern with policy and institutional coherence, underpinning all other thematic activities of the GFMD. More recently, under the guidance of the GFMD Working Group on Policy Coherence, Data and Research,[23] this concern has narrowed to impact evaluations as a basis for better understanding the connections between migration and development, and addressing them more coherently in future policies. For migration and development policy makers, and for the GFMD which seeks to bring the two together, this may be the most important challenge at the interface of migration and development today.

---

[23] See the description and rationale of the working group on the GFMD website: http://www.gfmd.org/en/adhoc-wg/policy-coherence-data-research.html.

GFMD 2010 in Puerto Vallarta gave special emphasis to impact assessments of migration and development programs and policies and the further development of Migration Profiles as an information tool to support these assessments. Koser traces the Global Forum's engagement with this issue since 2010, when a survey of GFMD governments on policy and institutional coherence was undertaken (following an earlier one in 2009), and a workshop was held in Vienna,[24] to the thematic meeting in Marseille in 2011 on assessing the impact of migration data and research tools to do this.[25]

The earlier surveys reveal that current practices diverge widely, and only a small number of mostly European countries have factored development impact assessments into their migration programs, and even those tend to be project specific. Most migration policies and programs are only evaluated for their outputs in line with the immediate project objectives. Since most countries do not mainstream migration in their national development plans, there is virtually no existing assessment work on either the impacts of migration or the impacts of the policies behind them. Questions remain about what kinds of data could satisfy both needs and what capacities are needed to collect and analyze data and utilize it for better linked-up policies (IOM 2010).

Policy evaluations can directly challenge government coherence, since they involve resources and funding across budget lines and coordination across a range of government and nongovernmental agencies. They raise the issue of available capacity, assessment mechanisms, and funding. They can be costly, and in the final analysis, many have only a low impact, or unintended consequences, and negative outcomes. Their call for a "culture of evaluation," political will, and adequate resources is especially challenging for low-income countries. And since the development aspect cuts across borders, they challenge governments to broaden their evaluation perspective beyond national impacts. The GFMD can take the lead by showcasing individual or joint evaluation efforts such as the 2011 survey by the UAE of the effects of temporary wage work in the UAE on Indian construction workers and their families back home, which indicated positive developmental impacts back home (see Textbox 5).

Koser suggests that, given the many players involved in migration, and the multiple impacts of migration, governments could undertake joint assessments of programs across borders.[26] There appear to be very few good practices in establishing the necessary institutional, bilateral, and multilateral frameworks for this. Another means of engaging governments at either end of the migration spectrum is to include migration in national Poverty Reduction Strategy Papers (PRSPs), which clearly are

[24] See the seminar on "Assessing the Impacts of Migration and Development Policies," Vienna, June 30–July 1, 2010, organized by the GFMD ad hoc Working Group on Policy Coherence, Data and Research (http://www.gfmd.org/en/adhoc-wg/policy-coherence-data-research.html).

[25] See the GFMD 2011 thematic meeting in Marseille, "Managing Migration for Development: Policymaking, Assessment and Evaluation," June 13–15, 2011 (http://www.gfmd.org/en/marseille-workshop.html).

[26] This was also discussed at the GFMD 2011 thematic meeting in Marseille.

linked to funding by interested donors, and subject to assessment of development relevance. PRSPs can link migration to strategies to reduce the pressure to migrate. But there is still insufficient systematic inclusion of migration in the PRSPs, and a large gap between intention and action. This is being addressed by the GMG agencies[27] that commenced work under GFMD 2011 on testing and piloting the *Handbook on Mainstreaming Migration into National Development Planning Tools* published jointly by IOM, ILO, UNDP, and UNICEF.

Among the newer information tools available to policy makers are Migration Profiles, which to date have gathered data on migration to assist with migration management. Originally conceived by the European Union, and taken forward in a number of countries around the world by IOM and the International Centre for Migration Policy Development (ICMPD), these are now also being adapted as tools for development planning. The European Commission and IOM are trying to standardize and globalize their application to this end.[28]

The GFMD has served as a useful vehicle for moving the international discussion and use of this tool forward in recent years. The GFMD initiatives in this field since Puerto Vallarta have also opened the way to fostering and strengthening policy evaluations among policy makers. Capacity building may be the biggest message out of the GFMD's engagement with impact assessments. Many governments face a serious gap between their assessment expertise and the more advanced knowledge of international agencies and expert groups.

In cooperation with IOM and other agencies, the GFMD has established a Migration Profiles Repository toward improving future data collection and assessment of the impacts of migration policies on development (http://www.gfmd.org/pfp/migration-profiles). Combined with a comprehensive note on how to implement Migration Profiles, this should offer useful models and guidance to interested governments in the future. From the broader perspective of institutional coherence, in some ways, the GFMD itself has become a cohering facility, as more and more governments bring their good practices and experiences to the table.

---

[27] The Global Migration Group (GMG) is an interagency group of UN organizations and commissions, and IOM, "bringing together heads of agencies to promote the wider application of relevant international and regional instruments and norms relating to migration, and to encourage the adoption of more coherent, comprehensive, and better coordinated approaches to the issue of international migration" (http://www.globalmigrationgroup.org/). It was formed in 2006 at the request of the then UN Secretary General, based on a recommendation by the Global Commission on International Migration (GCIM). It evolved from the smaller interagency group, the Geneva Migration Group, formed in 2003.

[28] Migration Profiles were first proposed by the European Commission in its Communication on Migration and Development in 2005 as a tool to gather data inter alia on labor markets in countries of origin and destination, migration and migrant remittance flows, and other issues relevant to migration, to help identify and develop strategies to address data gaps and produce the evidence required to inform policy [COM (2005) 390, p. 37]. The EC, IOM, and ICMPD have since assisted governments in a large range of countries to develop their migration profiles. The model has shifted and expanded more to become a tool for development planning in the future (http://publications.iom.int/bookstore/index.php?main_page=index&cPath=41_42).

Regional Consultative Processes on Migration (RCPs) and Interregional Fora (IRF)

RCPs and IRF count among the informal elements of "global governance," along with GFMD, the UN High-Level Dialogue (HLD), and the Global Migration Group (GMG), but operate at a more concrete, operational, outcomes-oriented level (IOM 2010). They have been a consistent focus of GFMD interest since 2007, because of the increasingly effective examples they offer of multilateral forums for cooperation and partnerships on managing migration to the benefit of lower-income countries.

GFMD 2010 in Mexico considered a new paradigm for examining RCPs and IRF and their contribution to strengthening the migration and development nexus, based on enhancing policy and institutional coherence. Capacity building was a keyword, since weak capacity in governance at national level can translate into weak capacity at the RCP and IRF levels.

On the margins of the GFMD, a number of cohering actions have been taken by and among RCPs to strengthen their interaction on migration and development issues. There is improved understanding of the role and functioning of RCPs and IRF as forums for dialogue and cooperation on migration, in part due to the GFMD preoccupation with these consultative forums.

Following an IOM-commissioned assessment of RCPs (Hansen 2010), their potential and actual impacts on migration governance are understood in a fundamentally new way. Their value is not principally in determining national policies, laws, and practice, but rather in complementing them. Their contributions are more wide-ranging and complex, related more to building trust between states, enhancing migration and development links, creating consensus, encouraging convergence, building capacities, and in these ways influencing change in institutional arrangements, laws, policies, and practice.

For Achieng, RCPs demonstrate how managing migration can be an exercise in voluntary "neighborhood" cooperation. RCPs are part of the larger economic or trade integration processes, but they have not become the same force for international cooperation and action as these processes. They are formal enough to offer a frame for bilateral cooperation (as in the case of the Social Security Agreements under the Mercosur[29] or Caricom[30] processes), but not binding enough to command and enforce multilateral cooperation. They serve some important practical purposes among neighbors, but may perhaps also delay any serious formal global system for managing migration.

---

[29] Mercosur (Mercado Común del Sur, or the Common Southern Market) is an economic and political agreement among Argentina, Brazil, Paraguay, and Uruguay, founded in 1991, to promote free trade and the movement of goods, people, and currency (http://www.mercosur.int/).

[30] The Caribbean Community (CARICOM), comprising 15 Caribbean nations and dependencies, promotes economic integration and cooperation among its members, inter alia by coordinating economic policies and development planning, and through special projects for the less-developed countries within its jurisdiction, and operating a regional single market for many of its members (www.caricom.org/).

RCPs contribute to capacity building, policy coherence, and interstate cooperation on migration. The GFMD has noted that the efforts of RCPs to increase interaction and concrete partnerships among countries may be one way of achieving enhanced cooperation on migration at the global level. While most regional processes on migration engage little with development issues, some are moving toward more linked-up migration and development policy making by increasingly connecting regions of migrant origin and destination. The Abu Dhabi Dialogue, for example, brings to the same table Colombo Process countries of origin from South and Southeast Asia and Gulf Cooperation Council (GCC) states of destination.[31] The joint experiments in protecting overseas contract workers in this context feature among the most progressive outcomes of GFMD-related initiatives in the past 2 years.[32] Similar recent exchanges between the IGC and APC processes, or the EU-Africa and Asia-EU dialogues, also promise gains from interregional exchanges.

These trends are likely to continue, and the GFMD will likely continue to speculate about whether these processes are offering models and elements for a more global migration management regime, or perhaps rather compensating for the lack of such a regime.

Civil Society, the Common Space, and the GFMD

Civil society and other nonstate agencies have been an indispensable part of the evolving global debates on migration and development, including in the GFMD. Yet their participation in the GFMD has been relatively restricted, in part due to the state-led nature of the Forum, in part also due to the difficulty of identifying and representing such a broad, disparate range of players. The GFMD in Puerto Vallarta cut through these complexities in 2010 by drawing civil society, international expert organizations, and governments onto the podium in a Common Space debate.

Chikezie reflects on where and how civil society, indeed nonstate agencies generally, fits into the government-dominated structure and culture of the GFMD. After 5 years, this remains a challenge, even a vexed issue, for the governments, civil society, and all other stakeholders in the process. Governments are generally keen to retain the state-led character of the GFMD. Yet many of the good practices considered in the GFMD have come from nongovernmental organizations, migrant and diaspora groups, or other expert agencies or have been taken forward by these partners as outcomes of the GFMD. Has the Puerto Vallarta innovation helped resolve this issue? The GFMD certainly remains one of the few, if only, global forums to seriously discuss government-civil society relations.

---

[31] See the description and rationale of the Abu Dhabi Dialogue at www.iom.int/jahia/Jahia/policy-research/.../abu-dhabi-dialogue.

[32] See also Chap. 2 by Philip Martin and the report on the GFMD 2011 thematic meeting held by the United Arab Emirates (UAE) in Dubai, January 18–19, 2011, on "Recruitment of Workers for Overseas Employment" (http://www.gfmd.org/en/recruitment-of-workers-for-overseas-employment.html).

Given the broad spectrum of "civil society" and the fact that nonstate players in the migration and development field extend far beyond the traditional concept of "civil society," Chikezie asks if it may not be time to speak of "nonstate partners" rather than strictly civil society. Chikezie sees this question as part of the bigger challenge of how to devise transparent, workable mechanisms to deal with migration and development in balanced ways, and to effect real change. The government-civil society issue is part of the "institution building" needed for global governance of migration and development.

Chikezie links the weak presence of civil society to the relatively weak treatment of development in the Forum. There has been a glaring absence of development agencies at the meetings. With the exception of a few countries like the UK, Sweden, and the USA (in 2010), few governments send development officials to the GFMD. The debates tend to be dominated by Migration, Interior, or Home Ministries, whose concerns are more with security and borders than development, or by Foreign Affairs Ministries, often more focused on process and strategy than on workable programs on the ground. Chikezie also sees the weak civil society inclusion as a reflection of the unclear, diffuse concept of "civil society."

At practical levels, a wide range of nonstate players deal with migration and development issues on a daily basis and can advise about what policies work on the ground and why they work. Diaspora businesses and enterprises experience the benefits of their investments in countries of origin when host governments offer them tax and other concessions for doing business abroad. Community-based organizations back home experience the direct impacts of lower remittances resulting from reduced salaries or loss of a job abroad. Grass-roots organizations see the negative social and economic effects of marginalization, exclusion from social support services, discrimination, and exploitation of migrants on their well-being and ability to support their families.

In 2010, the Mexican GFMD Chair-in-Office's introduction of the "Common Space" was intended to rectify what many governments and civil society members of the GFMD saw as the inadequacies of the "interface" between government and Civil Society Days (CSD) of the previous GFMD meetings. In 2011, the successor Chair, Switzerland, expanded this practice to its thematic meetings around the world, and in the concluding debate, and sought to correct the lack of engagement of the private sector, specifically business, at a conference in Berne on "Markets for Migration and Development (M4MD): Trade and Labour Mobility Linkages – Prospects for Development?" (www.gfmd.org/en/markets-for-migration-and-development.html). The Swiss Chair also broke with the tradition of the civil society setting its agenda strictly in line with that of the governments and left this to the civil society organizers in 2011.

Chikezie calls on the civil society and other nonstate agencies to recognize that the economic and employment challenges facing most countries today require a broader participation on the civil society/nonstate GFMD team. He specifically challenges the diaspora to engage more with the work of the Forum, a cry repeated at the 2011 thematic meeting in Berne on markets for migration and development.

This chapter does not propose an optimal institutional architecture for governance of global migration, but identifies steps (even if incremental) that key actors – such as GFMD Chairs, members of the GFMD Steering Group and Friends of the Forum, migrant/diaspora, development practitioners, the private sector, and organized civil society actors – could take to strengthen the partnership culture of the GFMD and the mutual benefits that the Forum can bring to all stakeholders. For example, broader engagement through multiple channels and stages might deliver more tangible results. Including civil society more directly in the agenda setting of the GFMD could also strengthen the development part of the Forum debate.

Setting a Global Agenda for Migration and Development?

There is growing speculation about whether and how the current international debate on migration and development, particularly in the context of the GFMD, may be strengthening multilateral cooperation. Newland poses the question in her chapter: Is the international debate leading to new frameworks that help states work together on global mobility issues, and encourage international organizations to pursue common agendas on migration? While the GFMD is a not a standard-setting process, it may become the forum that sets markers in key areas at the interface of migration and development. Though nonbinding, the Forum can promote cooperation and progress to test and operationalize these markers.

Discussing migration, a traditionally unilateral preoccupation, in the context of development, a global concern, has probably helped to promote global governance. The GFMD understands that unilateralism is inadequate in the face of globalized and interdependent forces such as labor mobility, trade, and development and focuses on collaboration and joint solutions. It does not generally speak of "global governance," but in fact what the Forum is attempting to achieve could fit Newland's definition: "global governance can be thought of as the set of collaborative practices that states and other actors adopt in order to bring about the best outcomes attainable and reduce the likelihood of the most feared outcomes materializing from global forces (such as migration)."

Newland suggests, however, that global governance may be more global in name than practice. Perhaps the GFMD is changing that by avoiding the stalemates or rancor experienced in other international forums on migration.

The GFMD has achieved a unique level of cooperation and trust on contentious issues such as irregular migration and family unity. A midterm internal overview of achievements of the GFMD observed in 2010 that while it is difficult to measure the outcomes, there have been many valuable outputs from both the roundtable discussions and the follow-up actions between GFMD meetings, ranging from practitioners' handbooks to pilot programs and impact assessments.[33] Other chapters

[33] This information is in part based on two internal surveys in 2009 and 2010 undertaken under the auspices of the Working Group on Policy Coherence, Data and Research but also on the follow-up actions under the Working Group on Protecting and Empowering Migrants for Development.

in this book also demonstrate a number of new laws and practices toward more development-inclusive migration policies. Institutional changes have also been spurred by the GFMD discussions, particularly in the area of national planning that links migration and development (Jamaica and Ghana).[34] The Forum has also developed a web-based platform to serve as a repository of GFMD outcomes and good practices, and as a vehicle to market and attract partnerships for programs and initiatives (www.gfmd.org/pfp).

Considerable progress has been made on several topics of interest to governments that were addressed repeatedly in GFMD roundtables. These include labor mobility, circular migration, temporary migration, diaspora engagement in the development of countries of origin, and migration data and research. The Forum's unique approach of joint preparations of roundtables by governments and other partners around issues of common interest has created what Newland describes as "communities of interest" that have together found greater clarity of concepts and principles.

As Newland observes, perhaps the greatest contribution of the Forum has been to build trust and develop habits of cooperation among governments, at least in the medium-to-long term. There have already evolved certain tracks of ongoing action (or plans of action) on some core themes that permit tracking progress toward some agreed goals. This may be the second most important achievement of the Forum in its brief history.

Can the GFMD strengthen multilateral cooperation on migration and development through new forms of state cooperation on global mobility and by promoting common agendas among international organizations? The GFMD has met the objectives set at the first UN High-Level Dialogue in 2006, but is unlikely to evolve into an overarching standard setter or invigilator of global migration governance. To underpin its accomplishments, more research is needed on the impacts of policies on migration and development, and of the GFMD on the formulation of policies on migration and development. Setting benchmarks would be useful to sustain cooperation among GFMD participants and strengthen the basis for migration and development policies.

Finally, international mobility is a journey of hope for millions seeking higher work and life opportunities. Such mobility is poised to grow, offering benefits for individuals and societies. The dialogue and collaboration fostered by the GFMD, the improved coherence fostered by discussions between governments and with civil society, and the best practices highlighted by governments have blazed a path to linking the relatively newer large-scale international migration of people with the longer-term concern for accelerating development.

The chapters in this volume show the potential of one of the world's newest international forums to better understand the linkages between migration and development and enhance the benefits of both through practical measures.

[34] These came to light during the thematic meetings on domestic workers in Kingston and Accra. Refer the respective Summary Reports for those meetings on the GFMD website.

References

Avendaño, A. (2009). *Reintegration and circular migration-effective for development?* Paper presented at the Session 2.2 Session 2.1, Roundtable 2: "Migrant integration, reintegration and circulation for development, Civil Society Days, Global Forum on Migration and Development", GFMD, Athens.

Commission of the European Communities. (2007). *Communication from the Commission to the Council and the European Parliament: On circular migration and mobility partnerships between the European Union and third countries*. COM(2007) 248 final, Brussels.

Global Forum on Migration and Development (GFMD). (2010, November 8–11). *Partnerships for migration and human development: Shared prosperity–shared responsibility: Report of the proceedings*. Fourth Meeting of the Global Forum on Migration and Development (GFMD), Puerto Vallarta, Mexico. Available at http://gfmd.org/en/documents-library/mexico-2010.html

Hansen, R. (2010). *An assessment of principal regional consultative processes on migration* (IOM Migration Research Series No. 38). Available from www.iom.int/rcp

Holzmann, R., & Pouget, Y. (2010, November 8–11). *Social protection for temporary migrant workers: Conceptual framework, country inventory, assessment and guidance*. A Study prepared for the Global Forum of Migration and Development (GFMD) Working Group on Protecting and Empowering Migrants for Development, in support of GFMD Puerto Vallarta, Mexico.

Martin, S. F. (2008). Research priorities and gaps: An agenda for the global forum on migration and development. In J. Chamie & L. Dall'Oglio (Eds.), *International migration and development: Continuing the dialogue: Legal and policy perspectives*. New York: Center for Migration Studies and International Organization for Migration at http://www.iom.int/jahia/webdav/site/myjahiasite/shared/shared/mainsite/published_docs/books/International_Migration_Development.pdf.

Martin, P. (2009, August). *Reducing the cost burden for migrant workers: A market-based approach*. www.gfmd.org

Chapter 2
Reducing Migration Costs and Maximizing Human Development

Philip Martin*

The Labor Recruitment Process

Labor recruitment aims to match workers with jobs. Job matching is a difficult and often costly challenge within borders (Barron et al. 1985) and is complicated by national borders that involve differences in laws and often language and culture between employers and workers.

The key ingredient for successful job matching is information. Information in the labor market is asymmetric. Employers know more about the requirements of the job they are offering than potential employees, and workers know more about their abilities than potential employers. Employers screen applicants to determine who is best qualified to fill a vacant job, and workers signal their abilities to employers by obtain degrees and certificates attesting to their knowledge and skills (Stiglitz 1975; Spence 1973).

There are three major ways in which information is transmitted between employers and workers—directly by employers advertising for workers and workers seeking jobs, via public employment services that post employer ads and advise job seekers, and via private recruiters that receive employer job vacancies and recruit workers to fill them. Within countries, most employers fill most jobs directly,

* This chapter was prepared by Professor Philip Martin, Department of Agriculture and Resource Economics, UC Davis, California, and elaborates on a paper prepared by the author for discussion at the Roundtable Session 2.1 on "Reducing the costs of migration and maximizing human development", GFMD, Athens, November 4–5, 2009 (Martin 2009): plmartin@ucdavis.edu

P. Martin (✉)
Department of Agriculture and Resource Economics,
University of California Davis, Davis, CA, USA
e-mail: plmartin@ucdavis.edu

I. Omelaniuk (ed.), *Global Perspectives on Migration and Development: GFMD Puerto Vallarta and Beyond*, Global Migration Issues 1, DOI 10.1007/978-94-007-4110-2_2,

meaning they advertise for workers online or in newspapers, hold or participate in job fairs that attract workers seeking jobs, and post job openings in the workplace and ask current workers to refer friends and relatives. Workers seeking jobs can apply directly with the employer after learning about the job.

Alternatively, employers seeking workers and workers seeking jobs can use the public or private job-matching services.[1] Some employers post job vacancies at public employment services (ES) or employment exchanges, where workers seeking jobs are usually interviewed by ES staff and referred to appropriate vacancies. Many ES agencies also distribute unemployment insurance (UI) benefits to jobless workers, enabling the agency to check whether jobs are available to those receiving benefits.[2]

The third major way to link employers offering jobs with workers seeking jobs is via nonprofit and for-profit private recruiters. For example, universities and other training institutions often maintain placement offices that allow employers to interview graduates seeking jobs. Private recruiters were once confined to the top rungs of the job ladder (headhunters seeking executives), but today, they match workers and jobs at all skill levels.

Recruiters can be information brokers, matching employers and workers, or they can provide additional services, from screening workers to supervising them at work. When a private firm recruits, deploys, and supervises workers at a workplace, it is normally considered a temporary help for employment agency. Under some countries' labor laws, an employment agency is considered the sole employer of the workers it brings to a workplace and supervises; in other cases, the agency is a joint employer with the firm benefiting from the work. Adecco, Manpower, and Kelly Services are examples of employment agencies that operate in many countries, usually as employers of the workers they place at a work site, although these multinationals rarely move low-skilled workers over national borders.[3]

[1] The basic job search model imagines that each job seeker has a reservation wage so that the duration of job search for a worker is a function of the frequency and distribution of job and wage offers received and the availability of unemployment insurance and other nonwork income. The length and cost of job search depend on when and how often applicants receive job offers, their variance (if most of the wage offers are similar, it makes little sense to wait for a much higher wage offer), and the cost of waiting for a better offer in terms of foregone income (Mortensen and Pissarides 1999).

[2] Public ES agencies can simply give all workers seeking jobs contact information for employers seeking workers, screen and test workers and only refer those who are deemed qualified, or arrange for the employer to meet job candidates at the ES office. In many countries, recipients of UI benefits who reject job offers can have their UI benefits reduced or eliminated.

[3] Workers placed at a firm by a third party generally earn lower wages than direct hires, reflecting in part the costs and profits of the employment agency. In the USA in 2002, median hourly earnings in occupations with both directly hired workers and workers hired via temp firms were as follows: direct-hire laborers, \$9.48 versus \$7.90 for laborers hired via employment agencies, helpers, \$9.25 versus \$7.79, and packers, \$8.03 versus \$7.65.

Job-Matching Recruiters

The International Labour Organization (ILO), which opposes labor being viewed as a commodity, has long seen private recruiters and employment agencies as potentially exploitative of workers. The ILO's first Recommendation in 1919 urged governments to

> prohibit the establishment of employment agencies which charge fees or which carry on their business for profit. Where such agencies already exist, it is further recommended that they be permitted to operate only under Government licenses, and that all practicable measures be taken to abolish such agencies as soon as possible. (Unemployment Recommendation, section I)

The 1919 Recommendation touched on international labor migration:

> the recruiting … of workers in one country with a view to their employment in another country should be permitted only by mutual agreement between the countries concerned and after consultation with employers and workers in each country in the industries concerned. (Unemployment Recommendation, section II)

Instead of private recruiters, the ILO Unemployment Convention No. 2 in 1919 recommended that governments "establish a system of free public employment agencies under the control of a central authority" (Article 2). The Fee-Charging Employment Agencies Convention No. 34 in 1933 (since superseded) called for abolishing fee-charging employment agencies within 3 years. This did not happen, and the Fee-Charging Employment Agencies Convention (Revised), No. 96 in 1949, repeated the call to abolish fee-charging employment agencies but included the option of allowing them to exist with appropriate government regulation. Convention No. 97 in 1949, which dealt with international migration for employment, stipulates "Each Member … undertakes to maintain… an adequate and free service to assist migrants for employment, and in particular to provide them with accurate information" (Article 2).

The ILO's Private Employment Agencies Convention (No. 181) in 1997 modified the 1919 Recommendations on private recruiters by acknowledging the useful role they could play in well-regulated labor markets. Convention No. 181 shifted the focus to regulating private recruiters, including the fees they charge to workers. Article 7 says that private employment agencies "shall not charge directly or indirectly, in whole or in part, any fees or costs to workers," although

> in the interest of the workers concerned, and after consulting the most representative organizations of employers and workers, the competent authority may authorize exceptions to the [no-fee] provision in respect of certain categories of workers as well as specified types of services provided by private employment agencies.

Article 8 re-emphasizes the importance of bilateral migration agreements: "Where workers are recruited in one country for work in another, the Members concerned shall consider concluding bilateral agreements to prevent abuses and fraudulent practices in recruitment, placement and employment."

The ILO, in advising countries how to implement Convention No. 181, emphasizes the importance of ensuring that recruiters provide all workers, and especially migrants, with a written contract in a language they understand that describes the job and spells out wages and benefits, hours and working conditions, and mechanisms for resolving disputes (2007, p. 26). The ILO also emphasizes the importance of requiring recruiters to disclose their fees fully to their employer and worker clients (2007, p. 29–31).

Principles 12 and 13 of the ILO's 2006 Multilateral Framework on Labor Migration call for "an orderly and equitable process of labor migration to guide men and women migrant workers through all stages of migration" by giving "due consideration to licensing and supervising recruitment and placement services for migrant workers" (p. 23–24). The guidelines to implement these principles call for licensing recruiters and respecting migrant rights with understandable and enforceable contracts. Guideline 13.7 maintains the bar on migrants paying fees; governments are to ensure "that fees or other charges for recruitment and placement are not borne directly or indirectly by migrant workers" (p. 25).

No-fee public ES agencies have a small and declining share of the job-matching market; in most countries, fewer than 5% of job matches are made via the public ES agencies. Larsen and Vesan (2010) attributed the problems of public ES agencies to asymmetric information. Public ES agencies, they say, wind up attracting job vacancies from the "worst" employers and deal with the "worst" workers seeking jobs, creating a "low-end equilibrium [that] is almost impossible to escape" (p. 1).

Public ES agencies have not fared well in the international labor market for two major reasons. First, there are few internationally recognized job descriptions and no global system to recognize and verify the credentials of workers gained outside the country of employment.[4] Second, employers often make a go-through-the-motions effort via public ES agencies to seek local workers before receiving permission to hire migrants. Most employers have identified the foreign workers they want to hire before seeking local workers via the ES, so their local recruitment efforts do not result in local hires.

Labor recruiters are in the information brokerage business (Stigler 1962). There are many information brokerage businesses that match buyers and sellers, from insurance services that match consumers seeking coverage to firms offering insurance to stockbrokers matching buyers and sellers of stock. However, labor recruitment has several unique attributes, including the nature of wage work and the exchange of effort for reward. The work transaction involves workers giving employers control over some of their time in exchange for wages. However, unlike buying and selling a good or service, placing a worker in a job at a particular wage is the first step in a continuous transaction, as employers monitor worker performance. Unlike most transactions for goods, where the purchase is made and the buyer and seller may never meet again, the employment relationship is continuous until terminated, either by the worker or the employer.

[4] There is an effort to standardize education in some fields, including accounting, IT, and health care, and to speed the assessment and recognition of credentials earned outside the country of employment, but this nascent training and job standardization is mostly confined to professionals arriving in some industrial countries.

Social networks are an alternative way to match workers and jobs. Currently employed workers know the requirements of the job and the capabilities of friends and relatives at home. Employers with "good" migrant workers often ask current workers to recruit friends and relatives when they need additional workers (Waldinger and Lichter 2003). Currently employed workers, in turn, recruit only those capable of doing the job and often take responsibility for orienting and training the workers they recommend to employers.[5]

Recruitment Costs

The motivation to cross national borders to work arises from the wage wedge or gap between earnings abroad and at home. The maximum recruitment fees that workers are willing to pay for jobs abroad depend primarily on the size of this wage gap. Other factors, including benefits, the nature of the work, and prospects for upward mobility, also play a role in a migrant's willingness to pay (Abella 2004).

For example, Bangladeshi migrants who expect to earn $25 a month at home and $250 a month abroad will not pay the entire $225 a month wage gap in recruitment fees. Instead, migrants typically expect to earn four to six times more abroad to compensate for separation from family and friends, suggesting they may pay up to half of their expected earnings to obtain a foreign job. Maximum recruitment fees are also influenced by the certainty of earnings at home and abroad. Indeed, the usual assumption of certain earnings at home and uncertain earnings abroad may be reversed if an informal worker in one country has a formal job and a contract abroad.[6]

The recruitment system, shaped by government rules and regulations in migrant-receiving and -sending countries, helps to determine how the wage wedge is divided between employers, who get work done cheaper with migrants than they would if they had to rely only on local workers, migrants who earn more abroad than at home, and recruiters and governments, who charge fees for migration-related services. The recruitment system has generally evolved in two major ways over the past four decades (Abella 1995). First, the worker-paid share of recruitment costs has risen over time, especially in Asia. Second, the share of recruitment costs borne

---

[5] Social networks are most common when much of the labor migration is unauthorized and informal. For example, unauthorized Mexicans often find US jobs via social networks, while recruiters are usually involved in filling jobs via US guest worker programs that admit low-skilled migrants, H-2A and H-2B (Massey et al. 1990).

[6] Harris and Todaro (1970) explained continued rural-urban migration to slums surrounding cities in developing countries by emphasizing expected wages. Migrants compared certain incomes in agriculture at home with the expected income from higher wages but also higher unemployment in cities, and moved if the expected urban income exceeded the anticipated farm income. Migrants moving from informal jobs at home to formal jobs abroad may move from uncertain to certain incomes.

by workers tends to rise as the skill level and earnings of migrant workers fall, primarily because the supply of potential workers rises faster than demand at the lower rungs of the job ladder.

The Four-Stage Recruitment Process

The international recruitment process begins with an employer in one country seeking workers to job vacancies. Information is the first economic cost of migration, and many so-called "bad" recruitment practices occur in this first information stage. Despite decades of migration experience, and the return of migrant workers from abroad, the young men and women who predominate among migrants often live in areas that do not have recruitment offices or the government agencies responsible for managing recruitment. This means that potential migrants usually rely on returned migrants and private recruiting agents and subagents to learn about foreign jobs.

Agents acting on behalf of recruiters are often the first point of contact for migrants seeking foreign jobs, and their compensation system may make them less than honest brokers. Most are residents of the area in which they recruit workers, enabling them to win the trust of local workers, but agents may not know or fully understand the requirements of the foreign job. Instead, the agent's incentive may be to get the worker to the recruiter's office in the capital city to sign a contract in order to earn a headhunter fee. The agent may also charge the migrant a fee for travel costs to the recruiter's office and additional fees as well.

First-level agents are very hard to regulate for several reasons. Some work for more than one recruiter. Even if first-level agents are employed directly by a recruiting agency, their activities may be hard to regulate because, for example, local-level financial transactions may be in cash and without receipts, making it harder to resolve disputes. Workers who have not been abroad tend to think of foreign earnings and local living costs, exaggerating the economic benefits of working abroad and encouraging them to pay higher fees.

During the second stage, migrants meet licensed recruiters with the goal of obtaining a contract for a foreign job. The major issues during this migrant-recruiter interaction are the wages, benefits, and working conditions spelled out in a contract, and the costs that the migrant must pay to get the contract, including travel and other costs. The third stage of the recruitment process involves obtaining the documents needed to travel abroad to work by both migrant-sending and -receiving governments, either at a one-stop shop or via multiple trips to multiple agencies. There may also be required predeparture training, health and skills certification, and other steps required to receive exit clearance.

The final stage of the recruitment process is leaving the country with an approved work contract, home-country passport and visa for work abroad, and airline ticket. Contracts and visas are often checked by sending-country government agency staff at departure airports. This exit procedure, which should be among the easiest to implement, is marred by frequent reports of escort services that enable departing migrants to avoid effective departure checks.

Recruitment Regulations and Realities

Asia, home to 60% of the world's people and 30% of the world's migrants, provides a laboratory to examine recruitment regulations and realities. The major migrant-receiving countries represent migration extremes. Japan has very few migrant workers, while Gulf oil-exporting countries rely on migrants to fill over 90% of private sector jobs. Singapore has an immigration policy that is most similar to non-Asian industrial countries, welcoming skilled foreigners to settle but rotating low-skilled foreigners in and out of the country. The Philippines is the major Asian labor sender, deploying almost 1.5 million workers a year abroad, while India and China are the leading recipients of remittances from citizens abroad.

All migrant-sending nations officially aim to lower the cost of recruitment. There are several reasons, including to enhance protections for low-skill citizens who earn critical foreign exchange and can enhance economic development. The examples in this section summarize recent migration patterns and efforts to improve the recruitment system to enhance worker protections as well as the development impacts of migration.

Bangladesh

Bangladeshi labor migration is large and was growing rapidly until 2008, when a record 875,000 Bangladeshis left to fill overseas jobs; the number fell to 358,200 in 2009 and rose slightly in 2010. Over half of Bangladeshi labor migrants went to Saudi Arabia between 1998 and 2004, but in 2008 and 2009, over half of Bangladeshi migrants went to the UAE. Even though the number of migrants deployed fell, remittances to Bangladesh rose to $11 billion in 2009 from $9.7 billion in 2008.

Many low-skilled migrants are in debt when they depart (Martin 2009). Predeparture migration costs are often twice the official maximum charge, which was 84,000 taka ($1,220) in 2009.[7] The men who predominate among Bangladeshi migrants pay higher recruitment costs than migrant women,[8] and many borrow some or all of the money needed to pay predeparture costs from private moneylenders. Interest rates can be 10% a month, so that a $2,000 debt can double in a year.[9]

---

[7] The 84,000 taka government-set maximum recruiting charge, effective in 2006, applies to migrants going to Gulf states and Malaysia. It was raised in stages from 8,000 taka in 1992 to 30,000 taka, 50,000 taka, 70,000 taka, and 80,000 taka. The maximum recruitment fee to send workers to Italy was set at 230,000 taka in 2002. Recruiters have generally charged migrants at least twice the official maximum recruitment charge (Islam 2009).

[8] Karim-Rajput (2011) estimated the average cost of migration at 106,000 taka for Bangladeshi women and 141,000 taka for Bangladeshi men in 2010.

[9] Rural moneylenders charge very high interest rates, often 60–100%. Migrants with few assets sometimes find a local guarantor to cosign the loan; the guarantor often receives 10% of the value of the loan for each year that a guarantee is provided.

Labor migration from Bangladesh is regulated under the Emigration Ordinance of 1982, which requires migrant workers to have valid contracts and work visas before leaving to fill overseas jobs. Section 8.1 allows the government to prohibit the out-migration of persons with particular skills "in the public interest," and allows Bangladeshis leaving illegally to be fined up to 5,000 taka or imprisoned up to a year. Section 10 authorizes the government to license recruiters and to withdraw recruitment licenses for violations of recruitment regulations, such as overcharging migrants (Section 23).

Three major agencies deal with the 3 Rs that link migration and development, recruitment, remittances, and returns:

- The umbrella Ministry of Expatriates' Welfare and Overseas Employment (MEWOE)
- The regulatory Bureau of Manpower Employment and Training (BMET)
- The government employment agency, the Bangladesh Overseas Employment and Services (BOESL)

The MEWOE, established December 20, 2001 under pressure from NGOs, has authority over all government agencies dealing with labor migration, including private recruiters and migrants, the labor attachés in Bangladeshi missions abroad, and the Wage Earners' Welfare Fund that assists migrants and their families. The MEWOE encourages the Bangladeshi Diaspora to invest in projects in their home communities and facilitates expatriate investments in Bangladesh.[10]

The Bangladesh Overseas Employment Policy, adopted in October 2006, promotes the out-migration of Bangladeshis and aims to protect them abroad. The Wage Earners' Welfare Fund was created in 1990 and, under December 2002 regulations, is supported by fees from migrants (initially 1,000 taka or $15), interest on the bonds posted by recruiters, and a 10% surcharge on fees charged by Bangladeshi missions abroad for passports and other services. The WEWF supports BMET-staffed desks at Zia International Airport that check and assist migrants leaving and returning to Bangladesh, including Probashi Channel buses that take migrants from the airport to the major Dhaka bus terminals for journeys to their villages of origin. The WEWF provides migrants abroad with legal help, transfers home the remains of migrants who die abroad, and provides support to the families of deceased migrants in Bangladesh as well as returning migrants to Bangladesh who were not paid by their foreign employers.[11] The WEWF is administered by a board that includes ministers of various government agencies and the recruiters' association BAIRA.

BMET, a government agency established in 1976 with 17 offices around Bangladesh, is charged with promoting the employment of Bangladeshis abroad

[10] Since 2008, Bangladeshis abroad who remit foreign currency to Bangladesh have "special citizens' privileges" in Bangladesh.

[11] Samren reported that the families of migrants who die abroad receive 100,000 taka ($1,450)—see www.samren.org/Govt_Programmes/bangladesh/B1.htm. In February 2006, it was reported that the WEWF death benefit was 100,000 taka, and that foreign employers usually pay 500,000 taka to the families of deceased migrants (Islam 2006).

and encouraging them to send remittances home. BMET also provides employment counseling to migrants, regulates private recruiters, and conducts research on migration and development. All Bangladeshi migrants must be registered with BMET; most recruiters help migrants to register with BMET as one of the final steps before their departure.[12] BMET reported in 2009 that some three million Bangladeshis have been registered since registration began in June 2004, including 2.6 million who left Bangladesh. Registered job seekers receive a BMET-issued photo ID that includes personal information.

BMET operates three 1-h predeparture training programs a day and issues the certificates of attendance (emigration clearance cards) required to leave Bangladesh—the cards are bar-coded and read by scanners in the departure terminal of the international airport.[13] BMET also handles complaints from migrants about recruiters, and returned 90 million taka ($1.3 million) to aggrieved migrants in 2008, when it reported receiving 392 complaints about recruiters in 2008.[14]

Recruiters point to the relatively few complaints to conclude that most migrants are satisfied with the recruitment system. Critics assert that it is hard for complaining migrants to win compensation, which discourages them from filing complaints. For example, Afsar (2009) reported that several migrants alleged that subagents and recruiters cheated them, but since they lacked receipts for the monies they said they paid, they were unable to win compensation (p. 50–51).

BOESL is a wholly owned government agency created in 1984. In 2009, BOESL had 38 employees and a budget of 20 million taka[15] ($290,000) to publicize the availability of Bangladeshi workers in foreign countries via its own offices and in collaboration with local partners. BOESL specializes in providing skilled and professional migrants for large civil engineering projects abroad, screening Bangladeshis interested in these foreign jobs, inviting foreign employers to Bangladesh to interview potential employees, and helping Bangladeshis selected to fill foreign jobs to obtain necessary documents and certificates. In most cases, foreign employers cover the predeparture costs of BOESL-deployed workers, although migrants going to Korea, most of whom are relatively low-skilled, must pay for Korean language training in order to have their names placed on the BOESL list from which Korean employers select migrants.[16]

[12] In an effort to avoid BMET registration as a last rather than a first step in migration abroad, BMET is supposed to operate a lottery so that a recruiter seeking 100 workers is provided with the names of 100 workers with the requisite qualifications who are registered with BMET. This apparently does not happen in practice.

[13] Afsar (2009) reported that only one of the 60 returned migrants attended the BMET predeparture training: "Interviews with the recruiting agencies suggest that they obtained BMET certificates on the respondents' [migrants'] behalf by offering extra money to the desk officer" (p. 24).

[14] The results of BMET inspections are not made public, although recruiters whose licenses are suspended are named.

[15] Six million taka, 30%, represents BOESL salary costs.

[16] BOESL noted that it must often cover at least the local costs of foreign employers who come to Dhaka to interview Bangladeshi workers.

Between 1999 and 2006, when private recruiters sent Bangladeshi workers to South Korea, the maximum recruitment fee was 210,000 taka. Under the Korea-Bangladesh labor migration memorandum of understanding (MOU) signed in 2006, only BOESL can recruit migrants to fill jobs under the Employment Permit System, and the maximum recruitment fee is 60,000 taka ($870).[17] BOESL sent 1,500 migrants to Korea and an additional 430 to other destinations in 2008, that is, the average cost of deploying 2,000 migrants was about $145, far less than typical private recruiter fees. Some 8,000 Bangladeshis passed the Korean language test,[18] which enables them to be placed on the recruitment lists from which Korean employers select migrants.[19]

According to government data, about 60% of Bangladeshi migrants leave on their own, 39% leave with the help of recruiters, and 1% leave via BOESL and other channels. However, most of the 60% who leave "on their own" in fact leave with the help of private recruiting agents who coach migrants to say they are leaving on their own.

Licensed recruiters have a monopoly on charging fees to Bangladeshis leaving to fill overseas jobs. There were 801 licensed recruiters in the Bangladeshi Association of International Recruiting Agencies (BAIRA) in June 2009 (831 in summer 2010), and 100 were reportedly active in deploying migrants.[20] Most recruiters receive 2-year licenses after submitting a police report attesting to their crime-free record and recommendations from two members of BAIRA. Recruiters must pay a 100,000 taka ($1,400) license fee and post a 1.5 million taka ($21,500) bond with the BMET (Ray et al. 2007, p. 32).

BAIRA, established with government support in 1984, aims to improve the image of the recruitment industry. BAIRA emphasizes the importance of migration to the Bangladeshi economy, noting that remittances exceed the net value of garment exports, and complains that the recruiting industry is taxed rather than the recipient of government subsidies, as with the garment industry. Recruiters are supposed to pay a 30% VAT tax on their revenues and must show they have paid this tax in order to renew their licenses; this tax may help to explain the paucity of receipts.

[17] Before the EPS system went into effect in 2004, four Bangladeshi recruiting agencies linked to Members of Parliament controlled the migration of trainees to Korea. They reportedly charged trainees 800,000–1,100,000 taka ($11,600–$14,500), which encouraged some to change employers to earn more, including traveling to Japan, where trainees were paid more. Korea eventually banned the recruitment of trainees in Bangladesh (Palma 2006).

[18] The Korean language test, which consists of 25 listening and 25 reading questions, is available at www.hrdkorea.or.kr/eng/main.html. Between 2005 and 2009, some 440,000 foreigners took the test, and about half passed.

[19] Migrants admitted under the EPS are entitled to the Korean minimum wage of 3,770 won ($3) an hour, but many, expecting to earn a million won a month with overtime, pay fees to recruiters anticipating higher-than-minimum wage salaries.

[20] The nonactive BAIRA member recruiters reportedly assist Bangladeshis who have received foreign job offers via friends and relatives abroad, or who are returning to foreign employers, to complete required paperwork in Bangladesh. These migrants go "on-their-own" in Bangladeshi migration data.

In addition, Bangladesh levies a 30% tax on airline tickets, which raises the cost of leaving the country.

The recruitment process begins with foreign employer demand letters or job offers. Most of the recruiters receiving these letters are based in Dhaka, most migrants are in villages, and the subagent Dalals make most of the first contacts with migrants.[21] Most migrants need the help of recruiters to obtain passports, visas, medical checks,[22] and air tickets, as well as clearance certificates from BMET.

The first point of contact for most migrants are the subagents (Dalals). Most are older, in their 1940s, and better educated than migrants, since they must be able to read work contracts and communicate with Dhaka-based recruiters—a few are returned migrants[23] (Afsar 2009, p. 18–19). Subagents explain the foreign job to often illiterate migrants and generally accompany migrants to Dhaka for passports and medical tests. Afsar (2009) reported that when interviewed, both subagents and migrants initially said they did not collect or pay fees, but probing questions revealed that subagents received commissions from Dhaka-based recruiters of 10,000–60,000 taka ($145–$870) per migrant. According to Afsar, migrants who trust village-based subagents do not compare fees charged by recruiters so that implementing rules to require recruiters and their subagents to advertise their charges may not reduce prices (p. 20).

The Bangladeshi government is often blamed for recruitment problems. In the Philippines, recruiting licenses cannot be sold or lent, and recruiters who do not deploy at least 100 migrants a year can lose their licenses; some BAIRA recruiters believe the Bangladeshi government should adopt similar regulations. The most common reform proposals involve substituting technology for recruiters, that is, using computers to record foreign job offers and to register potential migrants and making recruitment fees a share of expected foreign earnings instead of a fixed amount (Ray et al. 2007, p. 174–175).[24]

---

[21] Some Dhaka-based recruiters place ads in newspapers seeking workers to fill foreign jobs; some smaller recruiters are based outside Dhaka.

[22] Migrants going to the UAE must have medical checks done in a facility in Bangladesh approved by the GCC; these facilities are organized into GAMCA, the GAA Approved Medical Centers' Association.

[23] Most migrants expressed trust in the subagents, believing they could recover what they paid from the subagent if the foreign job did not materialize. In some cases, subagents were relatives of the migrants they helped to go abroad. Some subagents reported that if they did not trust the work contract, they would not give it to a relative or a person they knew; instead, they would only give uncertain contracts to strangers (Afsar 2009, p. 21).

[24] Ray, Kumar, and Chaudhuri make several sensible recommendations to improve recruitment, but the broader purpose of the 400-page report, to provide advice on new countries to which Bangladesh could send migrants, is flawed. The heart of the analysis is a list of countries and "labor shortages" by occupation (Table 5.5, p. 98–104).

It is not clear how the labor shortage lists were compiled. For the USA, the lists appear to be a mix of fast-growing occupations, such as nurses, and occupations projected to have shrinking employment, such as laborers. Instead of using official Bureau of Labor Statistics projections, the report relies on employer-advocate interpretations of them aimed at persuading the US Congress to approve new guest worker programs (p. 107–121). BLS did not project labor shortages, as the report asserts on p. 121.

Several foreign governments have halted the recruitment of Bangladeshi migrants because of forged skills certificates and high predeparture debts that encourage migrants to take second jobs or stay abroad. Without long-term relationships between Bangladeshi recruiters and particular foreign employers or recruiters, some may have an incentive to send unqualified Bangladeshis abroad. For example, an unskilled Bangladeshi may offer to pay more for a plumbing work contract promising $250 a month than for a laborer contract paying $175 a month. A Bangladeshi trained as a plumber may not accept the $250 wage, but the unskilled worker who pays more and goes abroad as a plumber may leave the foreign employer dissatisfied because he does not have plumbing skills.

India

India, with a Diaspora of at least 25 million, is the world's leading recipient of remittances, over $50 billion in 2009 compared to China's $48 billion. Even though media attention often focuses on Indian IT and health care professionals employed abroad, most Indian migrant workers fill low- and semi-skilled jobs in Gulf Cooperation Council countries (Khadria 2009).

Labor migration from India is governed by the Emigration Act of 1983 (as amended), which regulates Indian-based recruiters and establishes "protectors of emigrants" (POE) offices. However, 13 categories of persons are exempt from the requirement to obtain a POE emigration certificate requirement, including those with 10 or more years of schooling.[25] Some conclude that the POE process introduces another level of fees that migrants must pay while adding few protections (Rajan et al. 2008, p. 5).

There are four major channels through which Indians migrate abroad to work with a mandatory emigration check (ECR passports). Individuals can arrange foreign jobs via friends and relatives, foreign employers may recruit Indians in India, India-based firms with foreign projects can send their employees abroad, and Indian recruiters can match Indians with foreign jobs. There are also many unregulated subagents, both in India and abroad.

The 1983 Emigration Act (amended most recently in 2009) requires recruiters to obtain licenses that are normally valid for 3 years. Recruiters must post a bond based on the number of migrants they send abroad (Ministry of Overseas Indian Affairs, website). Registered recruitment agencies (RRAs) that send fewer than 300 workers abroad a year must post a bond of 300,000 rupees ($7,100 in 2008), those that send 301–1,000 migrants a year must post a 500,000 rupee bond, and those that

[25] India issues two types of passports—those for which an emigration check is required (ECR passports) by the POE and passports that do not require an emigration check (ENCR). Holders of "emigration-check-required" passports can nonetheless leave without a POE check for most countries, but need POE permission to travel to 17 countries, including the Gulf oil exporters and Malaysia and Libya.

send more than 1,000 migrants must post a one million rupee bond. Amendments to the 1983 Emigration Act in 2009 require a 2.5 million rupee bond good for 10 years.

Maximum recruitment fees were 2,000 rupees ($45) for low-skilled workers in 2008, 3,000 rupees for semi-skilled workers, and 5,000 rupees for skilled workers; amendments to the 1983 Emigration Act in 2009 substituted a maximum recruitment charge of 25,000 rupees or 45 days foreign wages, whichever is less. Many Indian migrants report paying far more. For example, an Indian gardener in Dubai reported paying $2,745 in 2008 to obtain a 3-year contract offering $136 a month for a 48-h week, suggesting that over half of his UAE earnings would repay recruitment fees (Davis 2009).

There were 1,835 RRAs in 2005, but over 4,300 RRA registration certificates had been issued (Rajan et al. 2008, p. 15). Recruiters pay service fees to the POEs that reflect the skill of the workers they send abroad. For instance, they pay 2,000 rupees ($47) to have the POE check the contract of a low-skilled worker, 3,000 rupees for semi-skilled workers, and 5,000 rupees for skilled workers.[26]

About two-thirds of Indian migrant workers leave from South India (20% each from Kerala and Tamil Nadu). Kerala is a state of 32 million with a per capita income of 63,000 rupees ($1,425) in 2009, the highest in India, as well as the highest Human Development Index score.[27] Kerala has a higher literacy rate than the rest of India, reflecting the decision of the Communist-led Kerala government to spend more than other Indian states on education and health, but is not considered friendly to business.

Even though half of Kerala residents are employed in agriculture and only 15% of Kerala households receive "household" remittances, Kerala is considered to be a migration-driven economy. Some 2.1 million Keralalites emigrated in 2008 (Zachariah and Rajan 2010, p. 5), when remittances of $5 billion were equivalent to 20% of Kerala's GDP. Almost 42% of Kerala migrants in 2008 were in the UAE. About a quarter of remittances to Kerala are spent on education, but educated residents who cannot find jobs locally often emigrate; unemployment in Kerala is almost 20%.

A survey of Kerala migrants found that 80% learned of foreign jobs from friends and relatives (Rajan et al. 2008, p. 24–5). The average cost of emigration was 57,000 rupees ($1,290), including $735 for the foreign work visa and $300 for air travel to the foreign destination (Zachariah and Rajan 2010, p. 40);[28] no migrants reported receiving bank or government assistance to cover recruitment costs. Migrants using recruitment agents had higher recruitment costs than migrants who used friends and

[26] In a bid to persuade foreign employers to use authorized Indian recruitment agents, the Indian government in spring 2008 began requiring foreign employers who do not use authorized recruitment agents to post a $2,500 bond, returned to the employer when the migrant returns to India, for each Indian migrant they recruit.

[27] Business Standard, June 28, 2010.

[28] Passports cost 1,170 rupees ($26) and emigration clearance 1,425 rupees ($32).

relatives to go abroad. Kickbacks on fees charged to migrants appear common—in some cases, recruiters get commissions from approved medical centers for each migrant they refer for the mandatory health check.

Reformers would like to eliminate emigration-check-required passports, make it illegal for Indian recruiters to charge fees to Indian migrants, and have the Indian government negotiate minimum wages for Indian migrants abroad. There are efforts to transmit electronically the key provisions of contracts signed in India to the UAE Department of Labor, which should help to resolve controversy over allegations of broken promises once migrants arrive in the UAE.

Nepal

Nepal is the poorest of the major South Asian countries sending workers abroad; the average per capita income of the 30 million residents was $400 in 2008 ($1,100 at PPP), according to the World Bank. Nepal's GDP expanded by 3.5% a year between 2000 and 2008, less than half the 8% growth rate of neighboring India, in part because of the disruptions and instability linked to a Maoist insurgency between 1996 and 2006. About 40% of Nepalese have incomes below the poverty line, and half of Nepalese families receive at least some remittances from abroad.

Peace in 2006 and rising remittances, $1.6 billion in 2008 and $2.1 billion in 2009, almost a quarter of GDP, contributed to recent economic growth.[29] Some 217,000 Nepalese migrants, mostly low-skilled and from rural areas of Nepal, left in FY 2008/2009, down from 249,000 in 2007/2008, but up sharply from 107,000 in 2003/2004. Most Nepalese migrants stay abroad several years. Riester (2009) put the stock of Nepalese working abroad at 1.6 million (1.7 million in 2010)[30] and reported that many Nepalese who lost their jobs during the 2008–2009 recession preferred to stay abroad and work in an irregular status rather than return to Nepal (p. 1).[31]

About 96% of Nepalese migrants going abroad in 2008/2009 were men, and many are employed abroad as security guards, drivers, or laborers. Apart from India, the leading destinations include Malaysia, which accepted almost 40% of Nepalese; Qatar, 25%; Saudi Arabia, 20%; and the UAE, 10% (Riester 2009, p. 2). There were an estimated 400,000 Nepalese migrants in Malaysia in summer 2010, when the Malaysian government halted the recruitment of additional Nepalese workers after a protest involving 5,000 Nepalese.[32]

---

[29] Seddon et al. (2002) argue that official Nepalese data on migrants and remittances understate both.

[30] Nepal News, August 10, 2010. Retrieved from http://www.nepalnews.com/main/index.php/business-a-economy/8242-nafea-asks-govt-to-restrict-individual-contract-for-gulf-countries.html

[31] Nepal News, September 6, 2009. Retrieved from http://www.nepalnews.com/main/index.php/business-a-economy/1300-overseas-jobs-getting-secure-for-nepali-workers.html

[32] Nepalese often work in global hotspots. One report in summer 2010 estimated that 300 Nepalese migrants a day were arriving in Iraq; others were arriving in Afghanistan to fill jobs.

In FY 2009/2010, about 80% of the 294,000 Nepalese who went abroad used recruiters; 20% left with individual contracts, and this 80/20 recruiter/individual contract ratio continued in the 2010/2011 fiscal year.[33] The Nepal Association of Foreign Employment Agencies (NAFEA) says that many Nepalese going to Gulf countries with individual contracts are headed for Iraq, and asked the government to halt individual contract migration to Gulf states. However, Sthaneshwor Devekota of the Foreign Employment Promotion Board (FEPB) said in October 2010 that the Nepalese government will permit Nepalese to migrate to Iraq, where wages are relatively high.[34]

Nepal, which has a labor force of about 12 million growing by 2.6% or 300,000 a year, has had an open border with India since 1950. This makes it hard to determine exactly how many Nepalese are leaving the country to work abroad and makes it difficult to protect migrants. In 2008, the Ministry of Labor and Transport Management estimated that 600 migrants a day left for non-Indian destinations and 400 a day for India. A large but unknown number of Nepalese are employed in Indian agriculture, construction, and services.

Nepal has a long tradition of migration abroad. However, there is a sharp contrast between the informal and low-cost migration of Nepalese to India and the much more formal and expensive migration of Nepalese to non-Indian destinations with the help of recruiters. Reports of abuse of Nepalese migrants abroad often blame Nepalese recruiters and recommend stricter regulation; they rarely consider alternatives. An increasing number of Nepalese women appear to be going to Gulf countries and Lebanon to work as domestic helpers, leaving from Indian airports with individual contracts.

The 650 members of the recruiters association NAFEA acknowledge problems in the recruitment system. Licensed recruiters are required to post a bond of Rs three million ($42,300), and the government establishes maximum recruitment fees. NAFEA says that potential migrants must spend Rs 10,000 ($140) to obtain a passport and make contact with recruiters and Rs 50,000–125,000 ($705–$1,760) to obtain a foreign job, but the government-set maximum recruitment fee for migrants headed to Gulf countries is NR 70,000 ($1,000).[35] Migrants are required to participate in an orientation session about their foreign job before departure, and the government in 2009 began to honor an outstanding recruiting agency each year.

A detailed Foreign Employment Act (FEA) was enacted in October 2007, followed by regulations in 2008 that, inter alia, grant 1-year renewable licenses to

---

[33] Nepal News, August 23, 2010. Retrieved from http://www.nepalnews.com/main/index.php/business-a-economy/8583-outflow-of-nepali-migrant-workers-declines.html

[34] On September 2, 2004, recruiting agencies in Katmandu were attacked after 12 Nepalese migrants were killed in Iraq. NAFEA continues to press the government for Rs 500 million in compensation for damage done to recruiter offices. Nepal News, August 10, 2010. Retrieved from http://www.nepalnews.com/main/index.php/business-a-economy/8242-nafea-asks-govt-to-restrict-individual-contract-for-gulf-countries.html

[35] Nepal News, October 20, 2010. Retrieved from http://www.nepalnews.com/main/index.php/business-a-economy/9940-agents-dismiss-govt-plan-to-issue-id-cards-for-migrant-workers.html

recruiting agencies and established a Department of Foreign Employment (DOFE) to administer the law. Chapter 9 of the FEA makes recruiters liable for penalties of 4–7 years imprisonment for making false promises to Nepalese going abroad or sending them abroad without proper documents. Aggrieved migrants can file complaints with the Foreign Employment Tribunal in Katmandu, where all government migration agencies and most recruiters are based. The FEA calls for the establishment of a migrant worker's welfare fund, labor desks at Nepalese airports to check the contracts of migrants leaving, and labor attachés abroad to assist Nepalese migrants in countries with at least 5,000 migrants (there are currently Nepalese labor attachés for migrants in Malaysia, Qatar, UAE, and Saudi Arabia).

The ILO noted that Nepal's FEA included so many "safe migration" provisions to protect women that, if implemented effectively, would make it almost impossible for recruiters to lawfully send Nepalese women abroad for fear of running afoul of the law's strong anti-trafficking provisions.[36] As a result, many Nepalese women migrate to foreign jobs via India, with few or no protections.[37] There are an estimated 114,000 Nepalese women employed in GCC countries, including 63,000 in Saudi Arabia.

The Nepalese government in 2009 had MOUs with four migrant-receiving governments, Bahrain, Qatar, UAE, and Korea. Of these, Nepalese migrants are most eager to work in Korea under the Employment Permit System, where they earn $1,000 a month or more in agriculture, fisheries, construction, small-scale manufacturing, and some services; Korea and Nepal signed an MOU on July 23, 2007. The number of Nepalese studying Korean and taking the Korean language test far exceeds the number selected to work in Korea. For example, over 36,000 Nepalese took the Korean language test in August 2010, and 4,200 passed, but Korean employers selected only about 2,700 Nepalese a year to work in Korea (Nepalese Government, EPS Korean Section, website).

Migration has a significant impact on the Nepalese economy and Nepalese villages, where peasant families sometimes borrow $2,000 to send husbands or sons abroad for 2 or 3 years to earn $200–$400 a month. However, there is little research on the development impacts of remittances and how they can be enhanced. The NIDS interviewed 60 returned migrants (2008, p. 16–18), two-thirds of whom had been abroad for less than 4 years, and found that almost 40% were unemployed after returning to Nepal. The government in 2009 planned to issue Rs seven billion

[36] The FEA did not reference Nepal's Decent Work Country Program for 2008–2010, which states "Given the lack of productive employment opportunities in Nepal, exacerbated by the decade-long conflict, foreign labor migration represents a possible channel for employing them effectively." Outcome 1 of the DWCP on "Improved policy coherence supporting increased productive employment opportunities for men and women" anticipated a national employment policy and "gender sensitive policies to protect migrant workers" by June 2009 (p. 7); this did not occur.

[37] Peace in 2006 brought a number of UN agencies to Nepal; coordination between them has been poor. UNIFEM, the major advisor on the FEA, did not consult the ILO; the Nepal UNDAF in 2008 referred only to UNIFEM in reference to migration. IOM negotiated a capacity-increasing agreement with the Ministry of Labor in 2008.

($92 million) in infrastructure development bonds to Nepalese migrants abroad, giving migrants an incentive to buy Nepalese bonds by allowing bondholders to remit other money to Nepal at no cost. This scheme did not work, but the National Planning Commission is developing a pilot program to offer loans to match with remittance savings to those organized into cooperatives to start small businesses.

Nepal developed its migration policies fairly recently with the assistance of international organizations that urged inclusion of the best-practice guidelines developed over the past quarter century. These best practices are reflected in the FEA and government announcements of protections for migrants, but the yawning gap between policy goals and realities has prompted the development of a new migration policy. It is not yet clear whether revising the 3-year-old FEA can narrow the gaps between the goals of Nepalese migration policies and the realities for migrants. Nepal remains a poor country with a tradition of out-migration, an open border with India, and relatively high recruitment fees.

Philippines

The Philippines is a major source of migrants, with 3.9 million Filipinos settled abroad in December 2008, 3.6 million legal temporary Overseas Foreign Workers, and 650,000 unauthorized migrants abroad (Commission on Filipinos Overseas, POEA, website). Remittances to the Philippines were $20 billion in 2009, over 10% of the country's $175 billion GDP.

Some 1.4 million Filipinos were deployed in 2009, up 15% from the 1.2 million deployed in 2008. Over half were rehires, meaning they had been employed abroad previously, and 330,000 were leaving to work on the world's ships. Of the 350,000 new hires in 2009, 93% went abroad with the help of private recruiters.

The Philippine Overseas Employment Administration (POEA) publishes data on the occupations of the 332,000 first-time OFWs (new hires). In 2009, over 40% were going abroad to be service workers and 35% to be production workers. The service workers presumably included 72,000 household service workers, almost all of whom were women.

Recruiting and deploying Filipino migrants is governed by the Migrant Workers and Overseas Filipinos Act of 1995 (RA 8042), as amended by the RA 10022. The two major government agencies involved are the POEA and the Overseas Workers Welfare Administration (OWWA). The POEA regulates recruitment in the Philippines and is responsible for labor attachés stationed at consulates abroad to provide assistance to migrants.[38] The OWWA provides predeparture orientation in the Philippines (since 2005) and has service centers abroad to handle emergency repatriation and provide various services to families left behind.[39]

[38] Filipino migrants have protested the charges levied by Philippine consulates for passports and employment contract authentication.

[39] The OWWA, funded by a $25 fee collected from migrants, provides emergency services at 28 Philippine embassies, including shelters abroad for migrants who run away from abusive employers.

These government agencies are financed largely by fees collected from migrants, including a P3,000 ($60) processing fee charged by the POEA, whose governing board includes representatives of the recruitment industry, and a $25 fee paid to the OWWA (Abella et al. 2004). The POEA reviews the contracts of migrants before departure and provides orientation to the countries in which they will be employed before issuing the Overseas Employment Certificates (OECs) necessary to leave the Philippines as a migrant worker.

Of the 332,000 land-based new hires in 2009, 53% were women. The largest single occupational category was household service workers, accounting for 22% of new hires (another 5% of new hires were caregivers and household-related service workers). Many Filipinos going abroad complete multiple 2- or 3-year contracts before returning to stay in the Philippines.

Recruitment agents are normally involved only in the first placement abroad; most migrants go abroad for second and subsequent trips without the help of a Filipino recruiter but may incur recruiter fees in countries of destination such as Taiwan. Recruitment-related fees, along with health and document fees and transportation costs, are the major expenses involved in working abroad. Filipino laws and regulations limit recruiter fees to a maximum 1 month's foreign earnings or 4.2% of expected foreign earnings for a 24-month contract. However, migrants who know that there are more applicants than foreign jobs sometimes pay more, often the equivalent of 2–4-months' wages to obtain the usual 2-year contract, making fees 8–16% of expected foreign earnings.[40]

Enterprises related to out-migration are a key part of the Filipino economy: labor recruiters, travel agents, remittance and insurance services are common, even in rural areas where the major link to the global economy is migration.[41] In October 2010, the POEA database included over 3,300 recruiting agencies, including 2,600 that sent workers to land-based jobs (some were delisted or suspended). There were about 1,200 licensed recruitment agencies active in 2004,[42] with combined annual revenues exceeding $400 million a year; Agunias (2010) reported 1,400 recruiters in good standing in 2010 (p. 22). Many recruitment agencies, which must be at least 70% Filipino-owned, belong to the Philippine Association of Service Exporters, Inc. (PASEI) or the Overseas Placement Agencies of the Philippines (OPAP).

[40] RA 1002 limits the interest on loans provided by recruiters to migrants to cover predeparture costs to 8%, and prohibits recruiters from requiring migrants to obtain loans from particular lenders. Violators can be subject to fines of P500,000, imprisonment of up to 6 years, and loss of their recruitment license.

[41] There are also a growing number of businesses that help migrants to invest their remittance savings at home. In 2004, there were over 200 cooperatives, associations, and self-help groups that pooled some of their remittances to launch small- and medium-sized enterprises. Many returned migrants buy in "jeepneys," the 20-passenger vehicles with facing rows of seats that transport riders for up to 4 km for 4 pesos ($0.08). A used jeepney can be bought for 250,000 pesos or about $4,500, can generate fares of 500 pesos a day, and after paying a driver and maintenance, will have paid for itself in a year or two. Other popular investments made by returned migrants are raising pigs or poultry or developing handicraft businesses.

[42] About 900 recruitment agencies specialize in placing Filipinos in land-based jobs abroad, and 300 in sea-based jobs.

The POEA campaigns against illegal recruitment and overcharging migrants, but there are almost daily press reports of migrants paying fees for nonexistent overseas jobs or paying a month's wages for a contract and waiting 4, 6, or 10 months before going abroad to earn wages (POEA, website). [43] Agunias (2010) reported that low-skilled Filipino migrants headed to the UAE for salaries of $432 a month paid $270–$400 in placement fees. It is not clear what these fees covered, presumably not travel costs, and the length of the contract was not stated, but a $400 fee should be within the 1-month's foreign earnings rule (p. 11).

POEA regulations require recruiters to obtain licenses and to post bonds, generally two million pesos ($44,000). Half of the recruiter's bond is a frozen bank deposit and half is a surety bond, and both are tapped to satisfy valid worker complaints of overcharging or unpaid wages. First licenses are probationary, and recruiters normally have to deploy at least 100 migrants a year to retain their licenses.[44]

The POEA has been handling a rising number of cases alleging illegal recruitment activities, up from 1,200 in 2003 to 1,600 in 2009, although the number of cases received has been declining, from about 865 to 250 between 2003 and 2009. The number of cases resolved has also been declining so that the backlog of cases has been rising to 1,400 in 2009. The number of persons arrested for recruitment violations has been rising and was 74 in 2009 (Table 18, POEA Statistics 2009).

POEA can close recruitment agencies that violate recruitment regulations. However, illegal recruitment is a criminal offense, so other government agencies must prosecute recruiters who could be fined and sentenced to prison. The National Bureau of Investigations attributes the relatively few convictions of recruiters charged with illegal recruitment to the inefficiency of the Philippine court system and the reluctance of many victims to file or pursue formal charges.[45] Some victims who file charges later refuse to testify because the recruiter is a relative, friend, or resident of their town. In other cases, victims who initially allege that they had to wait too long to be deployed are in fact sent abroad, which stops the case.

Many violations of POEA regulations occur overseas, as when a migrant is required to sign a supplemental contract that requires the payment of additional fees, the migrant's passport is held by an employer or recruiter abroad, or the migrant is not allowed to leave the country at the end of the contract. Migrants in such cases can complain to the labor attachés at Philippine consulates, and some do.

Filipino recruiters are jointly liable with foreign employers to fulfill the provisions of migrant contracts, a policy that should help to protect migrants. However, some recruiters complain that their revenues and profits depend on

[43] Most police stations in the Philippines have an OFW desk to receive complaints.

[44] The POEA deployed about 10,000 OFWs directly in 2003; the largest private recruitment agencies deployed 3,000–5,000 migrants.

[45] In November 1998, a Malaysian, a Briton, and an Australian were arrested after they recruited Filipinos for jobs in a fictitious country, the Dominion of Melchizedek, said to be in the South Pacific (in the Old Testament, Melchizedek was a model priest to whom Abraham and others paid tithes).

deploying migrants to employers abroad whom they may not know well, and some migrants who win cases against their recruiters are not compensated if recruiter bond funds are exhausted.

The Philippine labor migration system is often considered a model for regulating recruitment and protecting migrants abroad. The Philippines is widely acknowledged to have one of the world's most diversified portfolios of foreign labor markets for its workers, and most of its workers are deployed legally and employed legally abroad. The role of the government in regulating and protecting migrants is nonetheless controversial, with complaints ranging from excessive government and private fees levied on migrants to the government sending workers abroad rather than accelerating stay-at-home development. Many Filipino recruiters, as well as the Union of Filipino Overseas Contract Workers (OCW-Unifil), want less government regulation of recruitment, arguing that increasing regulation drives up the costs of sending migrants abroad at a time when other countries offering lower-wage workers are aggressively expanding efforts to promote their workers to foreign employers.

Sri Lanka

Sri Lanka is the richest South Asian country sending workers abroad; the World Bank reported that GNI per capita in 2008 was $1,800 ($4,500 at PPP). In 2008, some 252,000 Sri Lankans went abroad with contracts to fill jobs for 1 or more years; 93% went to Gulf-area countries. The Sri Lanka Foreign Employment Bureau (SLBFE) reported that 1.4 million Sri Lankans, about 7% of the population, had worked abroad between 1985 and 2010.

Between 2001 and 2005, annual deployments rose from about 184,000 to 231,000, about 25%. Some 247,100 Sri Lankans went abroad to work in 2009, down slightly from 250,000 in 2008. Over 80% of SL migrants in 2009 went to four countries: Saudi Arabia (32% of the total), Qatar (18%), Kuwait (17%), and UAE (17%).

Migration and remittances are central to the SL economy. SL had a labor force of 8.1 million in 2009; employment was 7.6 million. An estimated 1.8 million Sri Lankans were employed abroad, almost 20% of the total 9.4 million SL workers employed. Remittances rose from $1.2 billion to $1.9 billion between 2001 and 2005 and topped $3.3 billion in 2009, more than garment exports and more than twice the value of tea exports.

The single largest group of SL migrants are women going to Gulf Cooperation Council countries such as Saudi Arabia and the UAE to be domestic helpers. About 52% of SL migrants in 2009 were women, and 90% left to be domestic helpers abroad.[46] Since January 1, 2008, Sri Lankan domestic helpers in the UAE must be paid at least Dh 825 ($225) a month. The introduction of this minimum wage

[46] Another 59,400 departing migrants in 2008 were classified as low-skilled, while 2,800 were classified as professionals; almost all low-skilled and professional migrants were men.

prompted fears of fewer jobs, but the Sri Lankan consulate in Dubai reported that the number of SL domestic helper contracts rose from an average 650 a month in 2007 to 1,000 a month in 2008.[47]

Many of SL women going abroad to work are from rural areas of Sri Lanka; working in a foreign home is often their first wage work. The SL government's goal is to reduce the share of female domestic workers among Sri Lankan migrants from 50% to 25% and to better protect SL workers abroad. Departing migrants receive booklets in Sinhalese or Tamil with information on the destination country and SL labor attaché or welfare officer contact information. All migrant workers are required to complete a predeparture orientation of 12 days for most Gulf countries and 30 days for Singapore. Since 2009, departing female domestic workers must also participate in an 18-day predeparture English and literacy program.

In 2008–2009, the Sri Lankan government developed a labor migration policy with three major elements: better governance and regulation of labor migration, more effective protection of migrant workers, and enhanced development impacts of migration and remittances in Sri Lanka. Tripartite steering groups, chaired by the Ministry of Foreign Employment Promotion and Welfare, developed recommendations in changes in governance, protection, and development policies, which were incorporated into the National Labor Migration Policy for Sri Lanka (ILO Country Office for Sri Lanka and the Maldives, website) and accepted by the SL government in April 2009.[48]

The National Policy calls for migration to be mainstreamed into Sri Lanka's development planning, calls for sending more skilled workers to new destinations via better English-language training, and envisions improved training institutions that issue certificates of skill recognized in foreign labor markets. The policy also calls for a revised code of conduct for recruitment agencies and a model employment contract for migrant workers.

About 75% of SL migrants go abroad with the help of licensed recruiters. SL had 746 licensed recruiters in 2009, up from 626 in 2008.[49] All recruiters must be licensed by the Sri Lanka Bureau of Foreign Employment (SLBFE) and, until 2009, were required to belong to the Association of Licensed Employment Agencies (ALFEA); since 2009, membership in the ALFEA is voluntary, and membership dropped from 650 to 350. Licensed agents must post a cash bond with the SLBFE and provide a bank guarantee of 750,000 SL rupees ($7,000) that the SLBFE can tap if there is a valid worker complaint against the recruiter.

[47] Since August 2008, UAE households pay Dh 200 ($55) for an insurance policy to provide death and repatriation benefits for Sri Lankan domestic helpers; the insurance pays up to Dh 30,000 ($8,175) in death benefits.

[48] The governance part of the National Policy outlines the institutional structure to manage labor migration and the legislative and regulatory framework. The protection part covers predeparture, employment abroad, and return and reintegration. The development section outlines ways in which remittances and the return of skills can accelerate development.

[49] The largest recruiter in 2009 was Lord Manpower, which sent 10,600 migrants abroad, followed by Nafa Travels, 5,400 migrants, and Gulf Line, 3,100 migrants.

SL government oversight of foreign jobs and the recruitment process is extensive. Foreign employers must have their job offers approved by the SL mission in that country, with the job offer contract logged into the SLBFE database before UAE-based recruiters contact SL-based recruiters to lawfully seek SL workers to fill the jobs.[50] SL recruiters may place ads in newspapers to attract migrants, or use subagents to find workers in rural areas.

Once a SL worker is selected by a foreign employer or recruiter, SL-based recruiters normally help SL workers to obtain passports and visas, arrange for medical exams, and schedule the SLBFE's compulsory predeparture training course for first-time migrants. The final steps in the process involve the recruiter usually taking the worker's passport, the approved foreign job offer, and the worker's training or experience certificate to SLBFE, where the data are entered into a database that enables the SL mission in the country with the job to check and ensure that there are not several workers being recruited to fill one job.

Village-based subagents often match SL migrants with licensed recruiters. Subagents can charge high fees and make (oral) promises that are not met, as when they charge domestic helpers $200–$300 to guide them to a licensed recruiter. When migrants complain about fees or unmet promises, they have no proof that the licensed recruiter collected such fees or made these promises, making it difficult to resolve complaints. Between 1994 and 2006, the SLBFE received 101,000 complaints from migrants and settled 74,000 of them, paying $660,000 (Del Rosario 2008).

The SLBFE in 2011 began requiring subagents, who they named business promotion officers (BPOs), to register with SLBFE and prohibited BPOs from collecting fees from migrants. Furthermore, the SLBFE requires licensed recruiters to include recruitment fees in their ads and must get approval from SLBFE before running ads seeking SL workers to fill foreign job offers, effectively allowing SLBFE to regulate advertised recruitment fees.

The Sri Lanka Foreign Employment Agency (SLFEA) is a self-supporting government agency established in 1996 to place SL workers in foreign jobs. In recent years, the SLFEA has placed about 1,000 workers a year in foreign jobs, charging foreign employers and SL workers fees to support itself. Another alternative to private recruiters involves the International Organization for Migration, which is recruiting 500 Sri Lankan workers for a Brazilian company constructing an international airport in Tripoli; the SL migrants are to earn at least 58,000 SL rupees ($520) a month in Libya.

Sri Lanka created a Ministry of Foreign Employment Promotion and Welfare in 2007, abolished the MFEPW for about 6 months in 2010, and recreated the MFEPW in November 2010. As in other countries with a Ministry of Labor and another Ministry charged with improving wages and working conditions for migrants, there has been competition between the ministries and international organizations advising them. In Sri Lanka, IOM, an intergovernmental organization outside the UN

[50] The SLBFE plans to launch a Web-Based Recruitment System (WBRS) in 2011 that will allow foreign employers and recruiters to view the qualifications of SL workers registered with SLBFE.

system, is closely involved with the MFEPW; the ILO, a UN agency that involves employers, unions, and governments in policy development, is more closely involved with the Ministry of Labor.

Conclusions

There are significant potential benefits to lower migration costs for migrants, employers, and governments. Migrants can benefit from lower predeparture debts, more savings abroad, and an enhanced capacity to contribute to development while abroad and upon return. Foreign employers can benefit from more satisfied and productive migrant workers. Migrant-sending governments can benefit from more remittances and satisfied returned migrants who contribute more to development, while migrant-receiving governments can benefit from having to deal with fewer migrant workers who violate the terms of their visas by taking second jobs or overstaying.

Recruitment is one of the most difficult aspects of migration management, much tougher to regulate cooperatively than remittances. The shared goal of migrant-sending and -receiving governments after the September 11, 2001 terrorist attacks was to encourage migrants to send more remittances via regulated financial institutions to provide fewer opportunities for terrorists to use informal remittance channels to transfer funds over national borders. Formalizing remittance flows, as discussed at the GFMD, required lowering the cost of transferring small sums over borders, which was accomplished by (1) modifying identification rules to allow migrants to open accounts and send funds via banks and other regulated financial institutions, (2) educating migrants about the lower costs and security of regulated financial institutions,[51] and (3) taking advantage of technological innovations that reduce remittance costs, including transfers via mobile phones.

The financial benefits of reducing recruitment costs are potentially greater for a typical low-skilled migrant than the benefits of lower recruitment costs. A typical Bangladeshi pays $2,000 for a contract in Gulf oil exporters offering $200 a month, or $7,200 in earnings over 3 years.[52] If remittances are $5,000, reducing remittance costs from 10% to 5% saves the migrant $250, as remittance fees drop from $500 to $250. However, cutting recruitment costs in half could save the migrant $1,000, four times more.

Migrant-sending governments at GFMD meetings agreed that recruitment costs can be reduced by educating migrants, regulating recruiters, promoting competition

---

[51] Some migrant-receiving countries required employers of migrant workers to pay them via regulated financial institutions, such as requiring them to deposit migrant earnings in banks. The immediate goal of direct-deposit programs was often to expedite the resolution of unpaid wage disputes, but a side effect was to encourage migrants to use the bank receiving their earnings to remit funds to family at home.

[52] Sharma and Zaman (2009) put the average cost of migrating at $2,300 in 2007.

or ethical recruitment among recruiters, and establishing government recruiting monopolies. Education helps migrants to learn about foreign jobs and the proper procedures to fill them, which should minimize overpayments to recruiters that arise from lack of information. However, there are differences between education about remittances and recruitment that reflect the number of transactions, the regulatory framework, and technology. Jobs and workers are far less standardized than remittance transfers, the job-matching transaction occurs far less frequently, and there are often multiple considerations when employers evaluate migrants for ability and loyalty and migrants assess the wages, benefits, housing, and other aspects of foreign job offers. Recruitment transactions, in other words, are far more complex than remittance transactions, and the consequences of a "bad" job match are generally more severe than for a bad remittance transaction.

After education comes regulation. Most governments require recruiters to identify themselves by obtaining licenses, and many issue licenses only to recruiters who pass tests showing that they understand recruitment regulations and post bonds from which aggrieved workers are compensated. The Philippines makes local recruiters jointly liable with foreign employers to fulfill the terms of contracts that the employer, migrant, and recruiter sign before departure. However, even joint liability may not be a magic bullet to ensure the enforcement of recruitment regulations if migrants do not complain, or withdraw their complaints after recruiters make private settlements with complaining migrants.

Regulation is based on punishing recruiters who violate rules. The alternative to the stick is the carrot, offering benefits to recruiters who abide by government regulations and go beyond them to help match local workers with foreign jobs. Efforts to develop codes of conduct for recruiters, and to encourage recruiting associations to uphold them, may promote self regulation that benefits migrants and foreign employers. Such carrot incentives are more likely to have their desired goals if the reward for good recruiter behavior includes tangible advantages such as lower fees or faster processing.

Most economists believe that competition is the most effective market regulator. There are hundreds of private recruiters in countries that send migrants abroad, and some may engage in a form of cut-throat competition that winds up raising costs for migrants. If sending-country governments fostered competition between recruiters in ways that reward those who abide by recruitment regulations, the result could be an increased market share for the best recruiters. In markets from telecommunications to air travel, deregulation that promotes competition has increased innovation and quality and lowered prices. Promoting competition in ways that ensure basic safety standards are satisfied, as with air travel, could help to protect migrants and lower recruitment costs.

The final option is to have more government monopoly recruiting agencies. Many migrant-sending governments allow only one agency to send workers abroad under the terms of bilateral agreements or MOUs, as with the Korean Employment Permit System that recruits Asian workers to fill Korean jobs or the Seasonal Agricultural Worker Program that moves workers from Mexico to Canada. Government monopolies have the advantage of assigning clear responsibility for who should abide by recruitment regulations, but they may not be as flexible or cost efficient as private recruiters. Government agencies charging low recruitment fees may nonetheless be unable to do away with payments made by potential migrants before they are registered with the agency, that is, since most government recruitment agencies are in capital cities, migrants may pay intermediaries to learn how to access them.

The best single strategy for migrant-sending governments to reduce recruitment costs is to promote development. Migration costs can be reduced if workers are empowered to say no to excessive recruitment fees, which they are more likely to do if they have alternative earning opportunities at home. Similarly, foreign employers are more likely to abide by labor laws and regulations if they must do so in order to obtain the workers they want. For both workers and employers, economic incentives to comply with recruitment regulations are likely to be more powerful than the threat of enforcement in the ever-expanding international labor market.

References

Abella, M. (1995). Asian migrant and contract workers in the Middle East. In R. Cohen (Ed.), *The Cambridge survey of world migration* (pp. 418–423). Cambridge: Cambridge University Press.

Abella, M. (2004). The recruiter's share in labor migration. In M. Douglas & E. Taylor (Eds.), *International migration: Prospects and policies in a global market* (pp. 201–211). New York: Oxford University Press.

Abella, M., Martin, P., & Midgley, E. (2004). Best practices to manage migration: The Philippines. *International Migration Review, 38*(4), 1544–1559.

Afsar, R. (2009). Unraveling the vicious cycle of recruitment: labor migration from Bangladesh to the Gulf States. In *International Labor Organization* (Working Paper 63). Retrieved from http://www.ilo.org/declaration/whatsnew/lang--en/docName--WCMS_106536/index.htm

Agunias, D. (2010). *Migration's middlemen: Regulating recruitment agencies in the Philippines-United Arab Emirates corridor*. Retrieved from Migration Policy Institute website: http://www.migrationpolicy.org/news/2010_6_03.php

Barron, J., Bishop, J., & Dunkelberg, W. (1985). Employer search: The interviewing and hiring of new employees. *The Review of Economics and Statistics, 67*(1), 43–52.

Davis, U. C. (2009, April). UAE, Jordan. *Migration News 16*(2). http://migration.ucdavis.edu/mn/more.php?id=3515_0_5_0

Del Rosario, T. (2008). Best practices in social insurance for migrant workers: The case of Sri Lanka. In *ILO Asian regional program on governance of international labor migration* (Working Paper 12). Bangkok: International Labor Organization.

Harris, J., & Todaro, M. (1970). Migration, unemployment and development: A two-sector analysis. *American Economic Review, 60*(1), 126–142.

International Labour Organization. (2007). *Guide to private employment agencies – Regulation, monitoring and enforcement.* Retrieved from ILO website: http://www.ilo.org/sapfl/Informationresources/ILOPublications/lang--en/docName--WCMS_083275/index.htm

Islam, S. (2006, February 27).Short-term life insurance yoke on Bangladeshi migrant workers. *New Nation.*

Islam, N. (2009). *Skill enhancement for potential migrant and pre-departure orientation* [PowerPoint slides]. Dhaka: Bureau of Manpower, Employment and Training (BMET).

Karim-Rajput, A. (2011). *Towards a comprehensive recruitment monitoring policy in Bangladesh.* Dhaka: International Organization for Migration.

Khadria, B. (2009). *India migration report 2009: Past, present and the future outlook*. New Delhi: Jawaharlal Nehru University.

Larsen, C. A., & Vesan, P. (2010). *Why public employment services always fail*. Presented at the Annual ESPAnet conference in Budapest, 2010. Retrieved from ESPANET website: http://www.espanet2010.net/en

Martin, P. (2009, November 4–5). *Reducing the cost burden for migrant workers: A market-based approach*. Paper prepared for discussion at the GFMD, Athens.

Massey, D., Alarcon, R., Durand, J., & González, H. (1990). *Return to Aztlan: The social process of international migration from western Mexico*. Berkeley/Los Angeles: University of California.

Mortensen, D., & Pissarides, C. (1999). New developments in models of search in the labor market? In O. Ashenfelter & D. Card (Eds.), *Handbook of labor economics* (Vol. 3A, p. 1367). Amsterdam: North-Holland.

Nepal Institute of Development Studies. (2008, November). *A study of migrant recruitment practice in Nepal*. Prepared for the ILO/UNIFEM/EU Asian Program on the Governance of Labor Migration Program.

Palma, P. (2006, November 5). Manpower market shrunk for policy failure. *Daily Star*.

Philippines Overseas Employment Administration (POEA). (2009). *Overseas employment statistics*. http://www.poea.gov.ph/stats/2009_OFW%20Statistics.pdf

Rajan, S. I., Varghese, V. J., & Jayakumar, M. S. (2008, July). *Overseas recruitment practices in India*. Centre for Development Studies. Presented at the ILO meeting in Dhaka.

Ray, S., Sinha, A. K., & Chaudhuri, S. (2007, September). Making Bangladesh a leading manpower exporter: Chasing a dream of US$30 billion annual migrant remittances by 2015. Indian Institute of Management Calcutta. Prepared for Danish Embassy. www.ambdhaka.um.dk

Riester, A. (2009, September 28–29). *Impact of the global recession on international labor migration and remittances: Implications for poverty reduction and development in Nepal, the Philippines, Tajikistan and Uzbekistan*. Paper prepared for the conference "The Impact of the Global Economic Slowdown on Poverty and Sustainable Development in Asia and the Pacific", Hanoi. Retrieved from ADB website: http://www.adb.org

Seddon, D., Adhikari, J., & Gurung, G. (2002). Foreign labor migration and the remittance economy of Nepal. *Critical Asian Studies, 34*(1), 19–40.

Sharma, M., & Zaman, H. (2009, August). *Who migrates overseas and is it worth their while? An assessment of household survey data from Bangladesh*. World Bank Policy Research Working Paper WPS5018.

Spence, M. (1973). Job market signaling. *Quarterly Journal of Economics, 87*(3), 355–374.

Stigler, G. (1962). Information in the labor market. *The Journal of Political Economy, 70*(5) Part 2: Investment in Human Beings, 94–105.

Stiglitz, J. (1975). The theory of screening, education, and the distribution of income. *American Economic Review, 65*(3), 283–300.

Verite. (2005). *Protecting overseas workers: Research findings and strategic perspectives on labor protections for foreign contract workers in Asia and the Middle East*. Retrieved from Verite website: http://www.verite.org

Waldinger, R., & Lichter, M. (2003). *How the other half work: immigration and the Social Organization of Labor.* Los Angeles: UC Press. Available at www.ucpress.edu/book.php?isbn=9780520231627

Zachariah, K. C., & Rajan, S. I. (2010). Migration monitoring study 2008: Emigration and remittances in the context of the surge in oil prices. In *Centre for Development Studies* (Working Paper 424). Kerala: Centre for Development Studies. Available at http://www.cds.edu

Chapter 3
Textbox 1: Circular Migration as a Development Tool: The Mauritian Approach

Ali Mansoor, Anil K. Kokil, and Vivekanandsingh Joysuree

Circular migration has been at the cutting edge of the migration and development debate in the Global Forum on Migration and Development (GFMD) since its inception. It combines the interests of highly industrialized countries in meeting labor needs in flexible and orderly ways with the interests of developing countries in accessing global labor markets, fostering skills transfer, and mitigating the risks of brain drain. Economists agree that the potential developmental benefits accruing from more circular movement of workers globally can be enormous, particularly where policies protect and empower the migrants, diaspora, and their families.

Mauritius has been showcased in the GFMD as an example of how circular migration can be an integral part of national development planning. But Mauritius also shows how in a globalized labor market circular migration can bring lower and higher income countries to the table as equal business partners, with potential gains for both. What can we learn from Mauritius' experiments in this area?

Globalization: Reinventing Mauritius

Following major shocks to its economy, the Mauritian Government in 2006 embarked on a 10-year Economic Reform Program to better meet the challenges of

A. Mansoor (✉) • A.K. Kokil • V. Joysuree
Ministry of Finance and Economic Development, Port Louis, Mauritius
e-mail: amansoor@mail.gov.mu

I. Omelaniuk (ed.), *Global Perspectives on Migration and Development: GFMD Puerto Vallarta and Beyond*, Global Migration Issues 1, DOI 10.1007/978-94-007-4110-2_3,

globalization. The focus was on fiscal consolidation and discipline, improving the investment climate and turning Mauritius into one of the top 10 investment-friendly locations, mobilizing Foreign Direct Investment (FDI), and restructuring the economy to unlock economic growth and opportunities. One component of the Reform Program promotes the development of circular migration schemes with interested receiving countries in order to expand Mauritians' opportunities for skills enhancement abroad and new employment and business opportunities when they return home. Incentives for Mauritian diaspora to return or "circulate" their skills and other assets are an important facet of this circular migration approach.

Stepping into the World and into the Future

In an island economy undergoing economic transformation from a labor intensive to a service economy, there is little scope for low-skilled workers to find employment. Circular migration to countries in Europe, Australia, and Canada offers opportunities for the low-skilled, especially women, to gain international exposure and rapidly build up human capital and financial savings. These could be used upon return to start a small business and generally raise the financial, economic, and social status of the family back home. Bilateral agreements can enable the sending country to benefit from lower costs of legal entry and improved conditions for stay and return of migrant workers.

But other forms of circulation – of persons, skills, investments, and other assets – also occur among the more established diaspora abroad, where ties to the home community can have development outcomes. For countries like Mauritius, circular migration can facilitate greater integration into the global economy through the creation of transnational networks, investments, and improved flow of FDI. It can also foster the development of different sectors of the economy such as financial services, information and communications technology (ICT), and travel.

Mauritius' circular migration initiatives have the following development objectives:

1. Widen the scope of opportunities for citizens to take up employment abroad for a specific period, learn new skills, save part of their income before returning to the country to set up a small and medium enterprise (SME), or invest in other economic activities
2. Encourage remittance flows and enhance their development impact
3. Harness the intellectual and financial resources of the Mauritian diaspora

An Enabling Framework to Support Circular Migration

Institutional and policy coherence is pivotal to the success of Mauritius' circular migration endeavors. The reform program has focused public and private sector efforts on creating an investment-friendly business environment and enhancing human resource development. The National Empowerment Foundation (NEF) was set up as a fast-track public-private partnership to unlock opportunities for the unemployed and those affected by retrenchment, particularly women, through training and reskilling. The NEF has also been entrusted to implement circular migration programs. The Government through the Small and Medium Enterprise Development Authority (SMEDA) is providing several incentives for SME development, which include preparation of business plan, skills development, and marketing. The Enterprise Mauritius, another institutional arm of the Government, facilitates the marketing of SME products in the African region. As at date, the number of SMEs has increased from 75,000 in 2002 to 100,000 in 2010.

For circular migration to be effective, interested participants are requested to come up with a project idea before they embark on the program. They are then exposed to various business opportunities in the SME sector to better prepare for their return. In Canada, they are connected to relevant institutions responsible for SME development and are assisted in preparing the business plan.

The Government has sought creative ways to harness the intellectual and financial resources of its sizeable diaspora, estimated at more than 200,000 individuals. These include dual citizenship (1995) and no duties on cars brought back to Mauritius (2006). In 2009, the Board of Investment initiated the "Invest-Hotel" scheme to allow Mauritians abroad to directly participate as investors, entrepreneurs, and promoters in the tourism sector. All such initiatives encourage circulation between home and host countries, both by migrants and by diaspora.

Return and reintegration of migrant workers after spending time abroad is critical to circularity and its development potential back home. Incentives for migrants and diaspora include a matched savings scheme with the SME Partnership Fund on a 1:1 basis to help returnees double their savings for investment in small and medium enterprises (SMEs). The Government is expanding and modernizing the tertiary education sector; and vocational training institutions deliver courses and programs that are recognized internationally. All this helps to prepare workers to compete on the global markets but also attracts migrants and diaspora home again. It is worth noting that, with the several initiatives taken by Government, Mauritius has been able to improve its rating internationally and has been ranked first in Africa for the third consecutive time in a row in terms of overall Ease of Doing Business and

Good Governance. Moreover, Mauritius has recorded a huge increase in Foreign Direct Investment from Rs 2.8 billion (70 million euros) in 2005 to Rs 14 billion (348 million euros) in 2010.

Lowering the costs of migration is critical to the success of the circular migration program. Since the Mauritian Circular Migration Program is a Government-led initiative which is implemented by the NEF, all migrant training is free. The potential for high recruitment fees or abuse of migrants has also been avoided through the assistance of International Organization for Migration (IOM). A Circular Migration Database has been developed to reduce the costs to migrants by expediting recruitment via free online registration. The employers bear the cost of air tickets and initial costs on arrival (accommodation and transport), so ultimately the only costs to the migrants are visas and medicals.

The costs of training can also be carried by both the sending and receiving country. While the NEF and other Government institutions provide predeparture orientation sessions, basic ICT training, English courses on functional literacy, basic numeracy, and financial literacy courses, once in Canada the workers are able to upgrade their skills at Canadian training facilities. For a new batch of the workers expected to leave for Canada in 2011, Canadian employers are supporting the Mauritius Institute of Training and Development and bearing part of the training costs. Under the circular migration agreement with France, both countries bear the cost of training the workers in Mauritius.

Circular Migration with Canada

In March 2008, some 85 Mauritians were recruited to work for the Maple Leaf Fresh Food Processing Company in their pork meat and poultry processing plants in Canada. To date, some 300 workers have taken up employment in Canada on a pilot basis. This represents the first of several agreements to place Mauritian low-skilled workers in countries that need workers in sectors such as agriculture and home care.

The trilateral agreement involves the Mauritian Government, foreign employers (e.g., Maple Leaf), and IOM. It covers predeparture orientation, employment contracts, visas, travel arrangements, and language classes. It ensures migrants' rights are safeguarded and workers are protected from any form of exploitation. The agreement also covers postarrival assistance, stay and return issues for the migrants who come from a country with a proven history of a well-integrated diaspora. An IOM Liaison Office was set up in Mauritius to implement this program and other programs in the pipeline.

A monitoring and evaluation exercise carried out in August 2008 revealed that on average, the workers consistently remit some Rs 20,000 per month

(EURO513 at November 23, 2011). Some have had increases in salary due to satisfactory performance, and some have even received a promotion. Many use laptops to stay in touch with family members via Skype. Some have obtained a driving license and subsequently purchased a car in Canada. Meetings are frequently arranged with families of the migrant workers. It appears that some settle their debts quickly. The wives of some of the migrants are following training programs, which they were unable to do before.

Circular Migration with France

In September 2008, Mauritius concluded a Bilateral Agreement on Circular Migration with France, the first of its kind in the world. Under the agreement, Mauritian workers are allowed to work in France for a specific period, gain experience, and save money before returning to Mauritius to find employment in new sectors of the economy, set up their own business, or invest in other economic activities. It offers opportunities for some 1,000 Mauritians including students, professionals, and other workers (61 occupations) among others to be exposed to the world of work in France.

Both Mauritius and France will ensure that appropriate schemes are devised for the economic and social integration of the workers back in Mauritius. Returning workers who set up a business that leads to employment creation would receive from the French authorities appropriate training, equipment, and financial assistance for the launching and monitoring of their project. Such entrepreneurs would also be provided necessary assistance through SMEDA for the design of their business plan and related services to set up their SMEs. For this category, the French will put at the disposal of each beneficiary a maximum amount of 7,000 euros. However, this amount could be increased to 20,000 euros, if the beneficiary were to create at least five jobs.

Both Mauritius and France carry the burden of training. Some one million euros over a 3-year period has been placed at the disposal of Mauritius for training of trainers, a center to train professionals, and upgrading existing training facilities. Since ratification of the Bilateral Agreement on Circular Migration by France in April 2011, appropriate structures have been set up both in Mauritius and France to facilitate its implementation.

By taking these courageous and bold reform initiatives, Mauritius has emerged as the leading country in Africa and is poised to showcase its "Circular Migration Model" to the world in the firm belief that it could open new avenues for development for every country across the world.

Chapter 4
Textbox 2: Final Report from Sweden's Parliamentary Committee for Circular Migration and Development

Stephen Dippel*

In March 2011, Sweden's Parliamentary Committee for Circular Migration and Development, set up by the Government in 2009, presented its final report to the Swedish Minister for Migration. The report contains both proposals for legislative changes and other recommendations aimed at facilitating increased back-and-forth mobility between Sweden and migrants' countries of origin in order to promote its positive development effects.

The Committee defines circular migration as "temporary or more long-term and usually recurring legal movements between two countries. Movement can occur either through migrants moving from their countries of origin or residence to a host country and then returning, or through migrants moving to their countries of origin for a shorter or longer period." The Committee advocates the promotion of *spontaneous circular migration* – which takes place when there is an enabling legislative framework that facilitates mobility – in contrast to the managed form that takes place within the framework of specific programs and projects for temporary labor migration. The report was submitted to all concerned stakeholders for consideration and comments during the summer of 2011, and concerned ministries are now carefully considering the various proposals.

---

* Source: The entire report, which includes an English summary, is available on the Swedish Government website: http://www.regeringen.se/content/1/c6/16/51/52/8ab268ca.pdf

S. Dippel (✉)
Swedish Ministry of Justice, Stockholm, Sweden
e-mail: stephen.dippel@justice.ministry.se

I. Omelaniuk (ed.), *Global Perspectives on Migration and Development: GFMD Puerto Vallarta and Beyond*, Global Migration Issues 1, DOI 10.1007/978-94-007-4110-2_4,

The main recommendations and proposals include the following:

- A permanent residence permit in Sweden should be valid for 5 years after residency in Sweden has ceased, if the individual in question makes a special request to retain their permanent residency status.
- Rules regarding the reacquisition of Swedish citizenship should be made more lenient.
- It should be possible to grant time-limited work permits for a total period longer than 4 years if the migrant in question wishes to continue to work in Sweden temporarily, but does not wish to acquire permanent residency.
- The qualifying period for a permanent residence permit for persons who have held a temporary residence permit for work for a total of 4 years should be extended from 5 to 8 years in order to allow longer periods of absence from Sweden.
- Scholarships for international students from third countries should be increased.
- International students from third countries should be allowed to stay in Sweden for 6 months after completing at least two terms of studies in order to look for a job.
- Under certain conditions, it should be possible for unemployed persons who are entitled to Swedish unemployment benefits to apply for work in a third country while retaining their benefits for 3 months.
- A resource center for diaspora organizations should be established in order to support their development efforts in countries of origin.
- A website for comparing money transfer fees for remittances should be set up.

Chapter 5
Social Protection for Temporary Migrant Workers: What Programs Serve Them Best?

Robert Holzmann* and Yann Pouget

Introduction

Temporary, short-term, or circular labor migration[1] is gaining increasing attention in the global development discourse as a means of contributing to the development of (poorer, and often but not always demographically younger) sending countries while satisfying the labor force needs of the (often richer, but not always demographically older) receiving countries. "Temporary migrants" are here defined as foreign workers with a fixed-term labor contract and commitment to return to the sending country after expiration. For countries of origin and the migrants, such an approach promises access to work opportunities, human capital formation, and remittances

*This chapter was prepared by Robert Holzmann (World Bank and IZA) and Yann Pouget (World Bank, Marseille Mediterranean Center for Integration). It draws on key results of a study and report prepared for the GFMD Working Group on Protecting and Empowering Migrants for Development, in support of GFMD Puerto Vallarta, Mexico, November 8–11, 2010 (Holzmann and Pouget 2010a). The chapter has benefited from very useful comments and suggestions by Chukwu Emeka Chikezie, Liz Warn (IOM), and participants of the August 2011 Geneva workshop. The content presents the views of the authors and does not necessarily reflect the opinions either of the World Bank and its executive board or other institutions they are involved with. Contact address of the authors: Robert.Holzmann@iza.org

[1] There is no unique distinction between short-term, temporary, and circular migration in the migration literature. The terms "short-term" and "temporary" will be used interchangeably and do not exclude circular migration as repeated return for some of the temporary migrants.

R. Holzmann (✉)
World Bank, Washington, DC, USA

IZA (Institut fuer die Zukunft der Arbeit), Bonn, Germany
e-mail: RHolzmann@Worldbank.org; Robert.Holzmann@iza.org

Y. Pouget
World Bank, Washington, DC, USA

Marseille Mediterranean Center for Integration,
Marseille, France

I. Omelaniuk (ed.), *Global Perspectives on Migration and Development: GFMD Puerto Vallarta and Beyond*, Global Migration Issues 1, DOI 10.1007/978-94-007-4110-2_5,

without long-term brain drain. For many receiving countries, it promises to fill certain seasonal, temporary, or, perhaps, even permanent labor market needs across the skills spectrum at the lowest economic and political costs in terms of integration of new residents.

A critical dimension of temporary labor migration is the access to and scope of social protection benefits for migrant workers and their families in the receiving and/or sending countries[2]. "Social protection" is used in this chapter to mean public interventions that assist individuals, families, and communities to better manage (social) risks and to provide support to the critically poor (see Holzmann and Jørgensen 2000; World Bank 2001). Defined via instruments, these comprise social insurance (such as disability, unemployment, and old-age benefits) and social safety nets (such as social assistance benefits and public work programs) but also other risk management instruments that help individuals better manage risks (such as predeparture information and training, financial education, and access to financial instruments). Today, social protection is considered a critical element of sustainable and equitable development, individual welfare and global development effects depending upon design, scope, implementation, and policy and instruments financing. Many development economists and social protection experts suggest that temporary migrants should have the same access to social programs as workers in the receiving country. This is typically motivated by human rights considerations, the fear of discrimination and exploitation, and the assessment that temporary migrants have broadly the same needs and are exposed to the same risks as local workers.

Having potentially the same access to social protection as nationals is a luxury for temporary migrants to OECD and other higher income receiving countries such as Singapore and Hong Kong. In these countries, benefit coverage is mostly well above 90% of the population or universal for some benefits. Temporary migrants to low- and middle-income countries with benefit coverage for national workers in the low single or low double digits are likely to suffer the same fate as the majority of the native informal sector workers. A special situation are the Gulf Cooperation Council (GCC) countries, where the indigenous labor force has well-developed social protection programs that benefit from main government subsidies, while (temporary) migrant workers only have access to a limited set of social programs, including end-of-service pay.

The basic conclusion of this study is that most temporary migrants may be better served by over- as well as under-provisioning of social protection/social risk management programs in accordance with their specific risk profile and needs (i.e., the gap between their risk profile and the available informal and formal risk management instruments). The report derives this conclusion from a conceptual framework developed as part of the underlying study for this chapter, a review of an inventory of such benefits in a number of immigration-intensive countries, and an assessment of these benefits against the conceptual framework and lessons from case studies.

[2] For related papers addressing the issue of social protection for migrants see e.g. Ruhs 2002, 2005, Martin 2003, 2005, Abella 2006, ILO 2006, Koettl 2006, Agunias and Newland 2007, Vonk and Kapuy 2008, Avato et al. 2009, and Sabates-Wheeler 2009.

The research informing this chapter was initiated by the United Arab Emirates (UAE) as a contribution to the GFMD meeting in Mexico in November 2010. The purpose of the research was to identify good practices and provide conceptual guidance on the design and composition of appropriate benefit packages for temporary overseas contract workers. To this end, the GFMD report proceeds in three main sections: Section 1 provides a conceptual framework for social protection benefits of temporary migrants and suggests differentiating between three key types of migrant workers linked with the length of contract/level of skills. Section 2 presents an inventory of such benefits offered by nine OECD countries, Singapore, and six GCC countries based on a framework-derived template. Section 3 assesses these benefits against the conceptual framework, presents lessons learned from key case studies, and offers ideas on how to adjust end-of-service pay schemes to the need of temporary migrants.

This chapter summarizes the key findings of the GFMD research undertaken around (1) the conceptual framework, (2) a review of select program inventories, (3) selective case studies, (4) end-of-service pay and pensions, and (5) proposes next steps for policymakers. The annexes provide the summary tables of the inventory; for the detailed country tables, see Holzmann and Pouget 2010a.

The Conceptual Framework

The conceptual framework outlines five key considerations for developing and assessing social protection programs for temporary migrants:

1. Policy objectives: Any social benefit package will be a compromise between not fully consistent objectives of the key stakeholders, that is, sending and receiving countries (and different policy actors within), and the migrants. In order to achieve the best compromise benefit package, it is important that the objectives of temporary migrant programs with regard to economic development, labor market, and social protection are fully articulated by and known to each of the key stakeholders (see Holzmann and Pouget, 2010b). Triangulating these interests may not be simple and calls for a balanced approach under the following proposed policy objective: *To implement adequate social protection measures to reduce the vulnerability of migrants and their families at all stages of the migration cycle while avoiding benefit arbitrage and striving for neutrality in labor market participation and mobility decisions.*
2. First best consideration—potential full access to social protection as national workers**:** Temporary migrants should, in principle, have full access to social protection benefits and services as permanent migrant and national workers unless dictated by their special needs and circumstances. This nondiscrimination would be in line with human rights considerations and international law, as spearheaded by International Labour Organization (ILO) and United Nations (UN) conventions. This position would avoid ex ante discrimination and distortions in the labor market. However, equal mandated treatment under different needs, preferences, and circumstances may not be optimal for the welfare of the (temporary)

migrant worker. Any deviations from full access toward designing and implementing special, tailored benefit regimes for temporary migrants must take into account their special needs and an enabling environment in both the home and host country.

3. Second best considerations: A benefits package should take into account the special needs, preferences, and circumstances of temporary migrants in sending and receiving countries. Such "second best" considerations can be crucial to the delivery of welfare-enhancing social protection provisions. Temporary migration leads to new and special vulnerabilities for migrants. These need to be addressed through a number of specific interventions within the overall framework of managed migration between sending and receiving countries. Moreover, full coverage under existing benefit schemes may not necessarily be in the best interests of temporary migrants for a number of reasons:

   - Migrants may not become eligible for benefits due to design features (e.g., waiting rules), eligibility constraints rules (e.g., non-export of benefits), administrative setup, or simply lack of information.
   - Migrants may have access to some benefits through the sending countries and hence have no interest in double coverage and costs with and without double benefits.
   - Migrants are not interested in the full scope of benefits and the associated costs (contributions/lower net wage) and prefer a higher take-home pay. This may be due to the scheme's lack of credibility, an interest in self-management of risks by the extended family, or the availability of remittance-financed productive assets once returned, high discount rate, etc. The latter seems particularly important for long-term benefits related to distant retirement (i.e., old-age pensions and health care during retirement).

4. The scope and level of benefits should be linked to the length and type of contracts. This paper proposes to differentiate between three types of temporary migrants according to the length and type of contract (which often relates to the skill level): unskilled and low-skilled seasonal workers, with contracts below 1 year; low- to middle-skilled migrants with contracts up to a few years and limited expectations for renewal or residency; and middle- to high-skilled workers, with time-bound initial contracts as screening device and strong expectations for renewal/residency when successful. These three main categories of temporary migration have different implications in terms of risks and vulnerabilities and the implementation of social protection instruments to cover them. Moreover, the different types of temporary migration follow specific policy objectives for origin and destination countries, which helps determine who covers what.
5. Revised social benefit package for temporary migrant workers: Such a package could comprise two parts: (1) mandatory (minimum) benefits linked to labor-related short-term risks in the host country and (2) variable mandatory (minimum) plus cosponsored voluntary benefits linked to long-term risks in the home country.

   - Mandatory short-term benefits could include inter alia, work injury (accidents and professional diseases), basic health care benefits, sick pay, and maternity leave.

In principle and to avoid distortions on the domestic labor market, temporary migrants should be granted those short-term benefits by the host country on the same basis as national workers (with the exception of, perhaps, maternity leave and advanced medical treatment for seasonal workers). Unemployment benefits and access to active labor market services should be considered but will need to be determined by the type of labor contracts and in relation to termination benefits.

- The second part of the package depends upon the nature of the work contract and includes variable minimum mandatory work-related benefits and benefits for the family left behind, plus voluntary and cosponsored long-term benefits. The latter could include pre-saving arrangements for old-age and health care benefits for families left behind. Of course, for voluntary benefits, the sky is the limit. The proposed long-term benefits are closely linked to existing termination pay provisions.

Based on the above considerations, Table 5.1 offers a benchmark for social protection provisions in temporary labor migration schemes.

The Review of Country Inventories

The GFMD report provides comprehensive inventories of social benefit programs in nine OECD countries plus Singapore (referred to below as OECD+ countries) and six GCC countries, and further inventories focused on termination pay/end-of-service pay in these countries and in eight further countries in the Middle East and Northern Africa (see Holzmann and Pouget 2010a). The assessment of social protection benefits under the reviewed programs against the established benchmark is summarized in Tables 5.2, 5.3 and 5.4 and suggests, in general, broad compliance by most programs and most migrant-receiving countries. Increasingly, migrant-sending countries offer coverage for some basic and migration-related risks. The major findings of the assessment include:

- Short-term benefits (work injury, basic health care, maternity leave) are always provided, mostly by public mandate and provisions and nominally at par with national workers. This applies for all three types of migrant workers, which make these risk management instruments the core provision for the migrant workforce.
- Work-related benefits and services (ranging from predeparture training and non-discrimination legislation to unemployment and termination benefits) show significant variance across countries. For these benefits, major differences seem to exist between OECD+countries and GCC countries reflecting possibly differences in the importance of migration and hence migration management approaches. Migrants to GCC countries typically receive departure training, travel and housing provisions, and facilitated remittance transfers as well as severance/end-of-service pay. These provisions are offered by exception in the reviewed OECD+countries. The latter are legally strong on wage equality and

Table 5.1 Mapping benefits to labor contracts for temporary migrants

Type of benefits Type of contract	Short-term benefits	Work-related benefits	Family benefits*	Long-term benefits
Type 1: Seasonal labor migration schemes for low-skilled migrants	Full access to work injury benefits, basic health care benefits, and time-limited sick pay in the host country	Predeparture training/information Travel arrangements (arrival, departure) and housing provisions Repatriation insurance	[*Family typically left behind] Covered by home country's social protection programs	Covered by home country's social protection programs
Type 2: Nonseasonal and non-transitional temporary labor migration schemes for low- to medium-skilled workers in low-tech industries and services	Full access to work injury benefits, health care benefits, sick pay, and maternity leave in the host country on the same basis as national workers	Predeparture training/information on administrative and cultural issues Travel arrangements (arrival, departure, home leave) and housing provisions Access to finance and facilitation of remittances Wage equality with native workers/antidiscrimination policies Work-specific labor market integration and training Provisions in case of bankruptcy (insolvency fund) Repatriation insurance/cost coverage of emergency repatriation Severance pay/end-of-service benefits (unless access to pensions in home or host countries)	[*Family mostly left behind except in exceptional cases of allowed family unification] Health care, survivors benefits, and family benefits (provided by host or home countries)	Full access to and portability of pension (and health care) benefits with end-of-service benefits plus voluntary matching contribution schemes as second best

Type 3: Probationary labor migration schemes for middle- to high-skilled workers	Full access to work injury benefits, health care benefits, sick pay, and maternity leave in the host country on the same basis as national workers	Predeparture information and arrival facilitation Facilitation of remittances Access to the same work-related benefits as national workers	[*Family unification typically allowed] Health care and family benefits as national workers in host country with possible access in home country (for family members staying voluntarily behind)	Full access to and portability of pension (and health care) benefits

* offers an explanation for the application/restriction of the benefit
Source: Authors

Table 5.2 Social protection provisions for type 1 low-skilled seasonal migrants

	Short-term benefits	Work-related benefits	Family benefits*	Long-term benefits
Benchmark benefit package	Full access to work injury benefits, basic health care benefits, and time-limited sick pay in the host country	Pre-departure training/information Travel arrangements (arrival, departure) and housing provisions Repatriation insurance	[*Family typically left behind] Covered by home country's social protection programs	Covered by home country's social protection programs
Selected schemes in Australia, Canada, France, New Zealand, Spain, the UK, and the USA	The majority of countries under review provide full access to the short-term benefits package as in our benchmark. Notable exceptions include Australia (where health care benefits are covered by the worker and seasonal workers do not meet eligibility criteria for parental leave), the USA (where health care, sick pay, and parental leave are nonuniversal and employer-sponsored), and the UK (where seasonal migrants do not meet the eligibility criteria for work injury, sick pay, and parental leave)	Only Australia and New Zealand provide predeparture training and information in their seasonal migration schemes. Travel and housing provisions are covered by the majority of countries under review, except in France where it is non-compulsory and the UK where travel costs are covered by the workers. Only New Zealand provides repatriation insurance to its seasonal migrant workers. Only Spain and the UK provide unemployment insurance to seasonal migrant workers. Overall, the countries under review under-provide with respect to our benchmark	Only Spain and the UK provide family benefits to seasonal migrants but only in the host country. They are non-compulsory and uncommon in the USA and non-available in all the other schemes under review, mainly because of impossible family unification and coverage only in host country. The latter case is in line with our benchmark	In accordance with our benchmark, seasonal workers are not entitled to any long-term benefits in New Zealand and the USA. The other countries over-provide with regard to our benchmark in terms of access to pension benefits and portability of pension benefits
GCC countries	No special programs for seasonal workers	No special programs for seasonal workers	No special programs for seasonal workers	No special programs for seasonal workers

* offers an explanation for the application/restriction of the benefit

Table 5.3 Social protection provisions for type 2 low- to middle-skilled temporary migrants

	Short-term benefits	Work-related benefits	Family benefits*	Long-term benefits
Benchmark benefit package	Full access to work injury benefits, health care benefits, sick pay, and maternity leave in the host country on the same basis as national workers	Predeparture training/information Travel arrangements (arrival, departure, home leave) and housing provisions Access to finance and facilitation of remittances Wage equality with native workers/antidiscrimination policies Work-specific labor market integration and training Provisions in case of bankruptcy (insolvency fund) Repatriation insurance Severance pay/end-of-service benefits (unless access to pensions in home or host countries)	[*Family mostly left behind except in exceptional cases of allowed family unification] Health care, survivors benefits, and family benefits (provided by host or home countries)	Full access to and portability of pension and health care benefits with end-of-service benefits plus voluntary matching contribution schemes as second best
Selected schemes in Canada, Singapore, Spain, Switzerland, the UK, and the USA	The majority of countries under review provide full access to the short-term benefits package as in our benchmark. Only Singapore excludes parental leave. The USA uses a non-compulsory and employer-sponsored system (except for work injury insurance)	Predeparture training and facilitated remittance transfers are non-compulsory in all countries under review. Only Singapore and Spain cover travel and housing provisions and Canada only travel costs. Wage equality and antidiscrimination are guaranteed in all countries. Unemployment benefits are available in Canada, Spain, Switzerland, and the UK (if residence requirements are met). This is the only over-provision with respect to our benchmark. Severance pay is available in Canada, Spain, and the UK and non-compulsory/employer-sponsored in the USA. Provisions in case of bankruptcy exist in all countries except the USA and Singapore. For the rest, and with the exception of wage equality and antidiscrimination, the countries under review fall short of SP provisions for type 2 migrants with regard to our benchmark	Only Singapore does not provide family benefits to its type 2 migrants. As for most of the other SP provisions, it is non-compulsory and employer-sponsored in the USA	All countries under review except Singapore provide access to pension benefits and portability of pension benefits. Health care benefits after retirement are only granted in host countries if access to permanent residence

(continued)

Table 5.3 (continued)

	Short-term benefits	Work-related benefits	Family benefits*	Long-term benefits
GCC countries	In the majority of GCC countries, temporary migrants are entitled to health care benefits, work injury, sick pay, and maternity leave. Paternity leave only exists in Saudi Arabia. Bahrain stands as an exception for its non-compulsory health care and work injury system (employer-sponsored). Overall, this puts GCC countries in line with our benchmark	Predeparture training and information is non-compulsory but provided by major South Asian sending countries' governments through welfare funds. The same applies to facilitation of remittances. Travel and housing provisions are guaranteed in all GCC countries, except housing provisions in Bahrain. Wage equality with native workers is not guaranteed for temporary migrants in all GCC countries. Training is usually provided by the employer, except in Kuwait and Bahrain. Provisions in case of bankruptcy exist in Kuwait, Oman, and Bahrain. Repatriation insurance is available in Qatar, Saudi Arabia, and the UAE. End-of-service benefits are available in all GCC countries. Overall GCC countries fall short of SP provisions when compared to our benchmark for type 2 migrants	Family benefits are non-compulsory in all GCC countries (except health care in the UAE) and therefore matter of private agreement between the employer and his employee	Pension benefits are available for temporary migrants only in Saudi Arabia and the UAE. Pension benefits are non-compulsory in the other countries. In all countries, end-of-service pay is compulsory and substitutes for long-term benefits. Healthcare benefits after retirement are non-compulsory in all GCC countries

* offers an explanation for the application/restriction of the benefit

Table 5.4 Social protection provisions for type 3 middle- to high-skilled temporary migrants

	Short-term benefits	Work-related benefits	Family benefits*	Long-term benefits
Benchmark benefit package	Full access to work injury benefits, health care benefits, sick pay, and maternity leave in the host country on the same basis as national workers	Predeparture information and arrival facilitation Facilitation of remittances Access to the same work-related benefits as national workers	[*Family unification typically allowed] Health care and family benefits as national workers in host country with possible access in home country (for family members staying voluntarily behind)	Full access to and portability of pension and health care benefits
Selected Schemes in Australia, Denmark, France, New Zealand, Singapore, Switzerland, the UK, and the USA	The majority of countries under review provide full access to the short-term benefits package as in our bench-mark. A notable exception is Australia, where since September 2009, temporary migrants are responsible for health care costs for themselves and their families. The USA uses a non-compulsory and employer-sponsored system (except for work injury insurance)	Predeparture training and facilitated remittance transfers are non-compulsory in all countries under review. Unemployment benefits are only available in France, Switzerland, and the UK. For the rest, type 3 migrants are mostly treated as national workers in terms of work-related benefits in the countries under review. The latter is in line with our benchmark.	Only Australia and Singapore do not provide family benefits to their type 3 migrants. As for most of the other SP provisions, it is non-compulsory and employer-sponsored in the USA	All countries under review except Singapore provide access to pension benefits and export-ability of pension benefits. Health care benefits after retirement are only granted in host countries if access to permanent residence

(continued)

Table 5.4 (continued)

	Short-term benefits	Work-related benefits	Family benefits*	Long-term benefits
GCC countries	In the majority of GCC countries, temporary migrants are entitled to health care benefits, work injury, sick pay, and maternity leave. Paternity leave only exists in Saudi Arabia. Bahrain stands as an exception for its non-compulsory health care and work injury system (employer-sponsored). Overall, this puts GCC countries in line with our benchmark	Predeparture training and information is non-compulsory for temporary skilled migrants in GCC countries. The same applies to facilitation of remittances. End-of-service benefits are available in all GCC countries. Wage equality with native workers is not guaranteed for temporary migrants in all GCC countries, which does not put high-skilled migrants on an equal footing with native workers in terms of work-related benefits. GCC countries therefore under-provide with respect to our benchmark	Family benefits are non-compulsory in all GCC countries (except health care in the UAE) and therefore matter of private agreement between the employer and his employee	Pension benefits are available for temporary migrants (expats) only in the Kingdom of Bahrain. In all countries, end-of-service pay is compulsory and substitutes for long-term benefits. Health care benefits after retirement are non-compulsory in all GCC countries

* offers an explanation for the application/restriction of the benefit

antidiscrimination measures and selectively on unemployment benefits and provisions in case of bankruptcy. Unemployment benefits are available for non-seasonal workers in most reviewed OECD countries (and for seasonal workers in Spain and the UK) but only in one GGC country (Kingdom of Bahrain). Selective coverage exists for benefits in case of bankruptcy and repatriation in GCC countries. And severance pay provisions for nonseasonal workers that compensate for dismissal before the end-of-contract are the rule for OECD countries but do not exist for migrant workers in GCC countries. The reverse is true of end-of-service pay provisions that grant lump-sum seniority benefits. These are universal in GCC countries but are the exception in OECD countries. The unevenness of work-related benefits may be explained by local legal and policy environments that prescribe them in one case (e.g., housing and transport) but not in another (e.g., facilitation of access to finance and remittance transfers). This is an area where more analysis is needed. A special issue in GCC (and other) countries is the enforcement of due wages and benefits. Reportedly, firms do not comply with contractual obligations due to mere reneging or during the crisis because of insolvency, and migrants do not engage in legal actions because of fear, efforts, and costs even if such instruments exist. The actual scope of the problem is unknown, but apparently not negligible. For this reason, some countries (such as the UAE) have recently implemented a Wage Protection System that controls the payment of wages due, and there are plans to expand it to end-of-service pay.

- Migration-related risks for temporary migrants: Several major sending countries in South Asia have established migrant welfare funds and related emigration support services to protect their overseas workers (OSCE et al. 2007). These funds mainly provide emergency and repatriation services, work-related benefits and legal assistance, and welfare provisions for families left behind but typically exclude long-term benefits. Financing comes from compulsory workers' contributions fixed at about USD 25 per person, and the funds are administered by public or semipublic agencies. The Philippines have set up this welfare fund model, followed by Bangladesh, Pakistan, and Sri Lanka. India is in the process of creating one, and other sending countries in other global regions are interested in replicating this model. Despite certain differences in the governance structure and service delivery mechanisms, they share similar objectives. Strategies to cover risks and provide benefits and services can range from recruitment regulation, predeparture training and information, repatriation, and financing assistance to support families left behind. On paper, this is a very encouraging approach to address risks that can and will not be well covered by the receiving country. However, besides qualitative assessments on the Filipino experience (Ruiz and Agunias 2008) and on the Sri Lankan experience (Del Rosario 2008), no rigorous monitoring and evaluation of such migrant welfare funds have been performed.
- Family benefits (basic health care and family allowance), inasmuch as they are provided at all to national workers, are generally inaccessible to temporary migrant workers unless their families reside with them. Family unification is typically excluded for seasonal workforces and often not permitted under nonseasonal temporary work permits. For health care, this puts the onus on the

migrant-sending country and the need to organize coverage there. For family benefits, the high level of inter- and intragenerational redistribution required is apparently something that migrant-receiving countries are not willing to extend unconditionally to their temporary workforce.

- Long-term benefits: Coverage for pension benefits is provided in most OECD countries but in none of the GCC countries, where end-of-service pay constitutes a saving instrument that can potentially serve as a proxy for long-term benefits (retirement income and health care in retirement). In Singapore, migrant workers are excluded from the pension scheme unless they are permanent residents. In OECD countries, however, eligibility for pensions is typically conditioned on lengthy contribution periods or residency that is beyond the reach of some categories of temporary migrants: Seasonal migrants in New Zealand and the USA do not have access to pension benefits since they cannot have access to permanent residence and meet the eligibility criteria. Portability provisions depend on existing bilateral or multilateral social security agreements. In some cases like Australia, temporary migrants can claim their pension contributions upon departure and receive a lump sum. Health care benefits after retirement are typically granted in the host country only, which excludes most of temporary migrants who return home after their work period. Health care benefits are usually excluded from social security agreements, except in the case of reciprocal health agreements between Australia, New Zealand, and the UK, where eligible workers can have access to health benefits in one of these countries if covered in their home country.
- Differentiation: The country programs reviewed exhibit some expected differentiation between seasonal and nonseasonal migrants at the level of work-related benefits and pension benefits. However, there is little differentiation between the level of skills (low-middle and middle-high), that is, between type 2 and 3 of temporary migrant workers. This suggests that if countries are interested in a certain type of workers, it seems that the contract length matters more than the skill level.[3]

Selective Case Studies

The lessons from a few selected case studies from destination (Canada, New Zealand) and origin countries (the Philippines and Barbados) move the analysis a step further from the review of legal provisions by providing useful information on the

[3] Ruhs (2011) has identified a positive correlation between migrant rights in general and social rights in particular and the skill levels for all and temporary migrants (Figures 9 and 10). But he does not differentiate between the types of temporary migrant workers and their benefit programs. Hence, it is unclear if his finding would hold if he were to control for seasonal workers that are typically low-skilled. And the benefits he investigates are only a small subset of those for which temporary migrants may be eligible.

implementation of such legal provisions. With the notable exception of New Zealand, the scope of review and the quality of the available information call for caution, as hardly any of the programs have been rigorously evaluated for their intended impact. The main conclusions are the following:

- The common characteristics for the reviewed case studies are threefold: First, on paper and as presented, they look like promising good practices, and possibly replicable in other countries. Second, only very few of these examples have to our knowledge undergone a thorough ex post impact evaluation, and even less so has a rigorous evaluation been built-in ex ante as part of system design and implementation. Third, the available information for some examples from other sources suggests that they have not as yet fully delivered what was expected. This contrasts with the case of New Zealand's Recognized Seasonal Employers Scheme (RSE) with Pacific Island neighbors, which contains a monitoring and evaluation component that has allowed improvements in design and implementation of key policy features.

  Design issues refer mainly to potential contradictions in eligibility criteria for certain benefits and residence requirement or workers' characteristics: It typically happens when temporary migrants contribute to a social protection scheme from which they cannot benefit because their employment period is too short to meet the eligibility criteria. This can result in a net loss for migrant workers and insufficient coverage of key contingencies. Implementation issues relate to difficulties in access to social benefits for eligible claimants in the real world, mainly because of information issues, institutional context, or inadequate bureaucratic capacities and failing delivery mechanisms.
- Canada's Seasonal Agricultural Worker Program (SAWP) is considered a good practice by a receiving country, because it offers migrants the same rights to social protection as national workers, involves employers in designing and implementing the program, and under Canadian law gives migrants the same status as other expressly protected groups.

  The available information (patchy evidence and subjective assessment) by migration experts suggests that that equal treatment may not necessarily benefit the temporary migrants as they do not gain eligibility to some of the benefits or exhibit low demand for while participating fully in their financing. In particular, seasonal migrants contribute to the Canadian Employment Insurance program, which provides temporary financial assistance for unemployed workers (regular unemployment benefits) and those who cannot work for reasons of sickness (sickness benefits), childbirth (maternity benefits), parenting (parental benefits), or support to a family member gravely ill with a significant risk of death (compassionate care benefits).

  Despite their contribution to the scheme, seasonal workers cannot meet the eligibility criteria for the regular unemployment benefits since they can only stay in Canada if employed and may not work enough to meet the eligibility requirements for the other benefits. This example demonstrates a fundamental design inconsistency.

- New Zealand's Recognized Seasonal Employers Scheme (RSE) was introduced in 2006 through the coordination among ministries and development agencies in New Zealand and in collaboration with ministries of labor and community leaders in Pacific Island countries involved in the selection of potential seasonal migrant workers. It aims to create a mutually beneficial circular migration scheme with access to secure labor supply for employers in the New Zealand horticulture and viticulture industries, while selected Pacific Island workers can secure access to the New Zealand labor market and return with experience and remittances. Table 5.5 provides a summary of the benefits for these temporary (type 1) migrants. A special feature of this scheme is the thorough monitoring and evaluation component that was built ex ante into the program design with the assistance of the World Bank. As a result of the monitoring and evaluation effort carried out by New Zealand's government, the design and implementation of certain work-related benefits in the RSE have been significantly improved (Luthria 2011). For example, predeparture training in the first season was not seen as successful as intended by employers and government officials, which spurred noticeable changes for the second season. The predeparture briefings have been adjusted to better meet worker's needs, notably with increased emphasis on worker education in sound budgeting and financial decisions (financial literacy and education). Another interesting development happened in the area of saving mechanisms: Since workers failed to save enough in the first season and expressed interest in saving more in the second season, several employers set up voluntary saving schemes with access to a bank account and remittance transfer facilities (Table 5.6).
- The Philippine Overseas Workers Welfare Administration (OWWA) has been managing a welfare fund of US $172 million in 2005 and had over 1 million members in 2007. It is fully funded by a mandatory membership fee of US $25 per contract for temporary migrant workers. This membership fee is mostly paid by the migrant but can also be paid by the employer. OWWA provides a broad range of services from regulation of recruitment, predeparture training and information, repatriation, and financing assistance to support families left behind. While the Fund exhibits all elements of a good practice program, the available assessment by Agunias and Ruiz (2007) highlights certain limitations in actual service delivery. A particular concern is the relatively low number of insurance claims made by eligible migrant workers, which shows that they do not make full use of their welfare entitlements. There is insufficient information to determine the reasons for this low insurance claim level. It is not clear if it is due to design issues or implementation issues. Only rigorous monitoring and evaluation would furnish the necessary information to improve this policy component.
- Barbados first extended social security to migrant workers with the Caribbean Community (CARICOM) Reciprocal Agreement in 1996. This totalization agreement involved 16 states and allowed legal migrant workers to qualify for maximum pension benefits and to claim other safety-net allowances, such as workers' compensation and unemployment benefit. A first assessment of the CARICOM agreement suggests that very few eligible workers have claimed their

entitlements. While nothing official has been published (even less an evaluation undertaken), direct information by the social security authorities suggests that the small number of applications is potentially due to design inconsistencies and lack of public awareness (Forteza 2010).

A Special Concern: End-of-Service Pay

A special concern for temporary migrants is their access to long-term benefits, in particular retirement income and health care benefits after retirement, in view of their temporary and often unpredictable migration cycle. The review of country inventories and special reflections on end-of-service pay schemes yields the following conclusions:

- The provisions in OECD countries (not Singapore) provide coverage of old-age benefits as migrants as other workers need to contribute to a social insurance scheme or are covered under universal provisions. But this does not make them necessarily eligible for these benefits when they retire or leave the country. Eligibility is often linked to a minimum number of years of contribution or residency that temporary migrants typically do not fulfill, and portability of acquired rights is more the exception than the rule. Hence, in the absence of effective bi- or multilateral agreements that seek to overcome some of these constraints, temporary workers risk paying contributions for a benefit they can never enjoy. This makes pension contributions equivalent to a wage tax that is likely to reduce the take-home wage of the temporary migrant and create a fiscal benefit of the receiving country.
- Addressing these deficiencies is limited to a few options, essentially two main sets: (1) unilateral or bilateral actions ranging from reimbursement of contributions paid (own and employer's share) on departure to effective agreements between social insurance institutions in receiving and sending countries, including totalization (i.e., the benefits are calculated on the total contributions paid independently), and (2) a change in benefit type away from the typical defined benefit schemes toward (funded or notional) defined contribution schemes that allow portability with and without bilateral agreements (see Holzmann and Koettl 2011).
- The GCC countries do not provide pension benefits for temporary migrants (expatriates), but all have end-of-service benefits (EoSB) arrangements amounting to about 1 month's wage for each year worked. These EoSB arrangements are akin but not fully equivalent to a defined contribution scheme as both provisions present different risk profiles and different requirements on the financial sector. Most importantly, EoSB run a major compliance risk by the employer who may renege on the obligation due to unwillingness, illiquidity, or insolvency. The equivalent counterpart risk with funded provisions is likely to be smaller but the investment risk larger, in particular if the funded provisions take

Table 5.5 Summary of inventory of social protection in selected temporary migration schemes – 9 OECD

Scheme type	Country	Short-term benefits: Health care benefits	Work injury benefits/ sick pay	Maternity leave/ paternity leave	Work-related benefits: Predeparture info and training	Travel provisions/ housing provisions	Emergency repatriation insurance/ cost coverage	Access to finance/ facilitated remittance transfers
Type 1	Australia	No	Yes/yes	No/no	Yes	Yes/yes	NCU	NC/yes
	Canada	Yes	Yes/yes	Yes/yes	NCU	Yes/yes	NCU	NC/NC
	France	Yes	Yes/yes	Yes/yes	NCU	NC/NC	NCU	NC/NC
	New Zealand	Yes	Yes/yes	Yes/no	Yes	Yes/yes	Yes	Yes/yes
	Spain	Yes	Yes/yes	Yes/yes	NCU	Yes/yes	NCU	NC/NC
	United Kingdom	Yes	No/no	No/no	NCU	No/yes	NCU	NC/NC
	United States	NC	Yes/NC	NC/NC	NCU	Yes/yes	NCU	NC/NC
Type 2	Canada	Yes	Yes/yes	Yes/yes	NCU	Yes/NC	NCU	NC/NC
	Singapore	Yes	Yes/yes	No/no	NCU	Yes/yes	NCU	NC/NC
	Spain	Yes	Yes/yes	Yes/yes	NCU	Yes/yes	NCU	NC/NC
	Switzerland	Yes	Yes/yes	Yes/NC	NCU	NC/NC	NCU	NC/NC
	United Kingdom	Yes	Yes/yes	Yes/yes	NCU	NC/NC	NCU	NC/NC
	United States	NC	Yes/NC	NC/NC	NCU	NC/NC	NCU	NC/NC
Type 3	Australia	No	Yes/yes	Yes/yes	NCU	Yes/NC	NCU	NC/NC
	Denmark	Yes	Yes/yes	Yes/yes	NCU	NC/NC	NCU	NC/NC
	France	Yes	Yes/yes	Yes/yes	NCU	NC/NC	NCU	NC/NC
	New Zealand	Yes	Yes/yes	Yes/yes	NCU	NC/NC	NCU	NC/NC
	Singapore	Yes	Yes/yes	Yes/no	NCU	Yes/NC	NCU	NC/NC
	Switzerland	Yes	Yes/yes	Yes/NC	NCU	NC/NC	NCU	NC/NC
	United Kingdom	Yes	Yes/yes	Yes/yes	NCU	NC/NC	NCU	NC/NC
	United States	NC	Yes/NC	NC/NC	NCU	NC/NC	NCU	NC/NC

Acronyms

- Yes means "legally provided in the receiving country (statutory) or specified in a labor agreement or memoranda of understanding (MoU) between the receiving and sending countries involved"
- NC means "non-compulsory." It is used when the social provision under review is not required by law in the receiving country (non statutory) or not specified in labor agreements or MoUs between the receiving and sending countries involved. It typically leaves the room for employer-sponsored benefits
- NCU means "non-compulsory and uncommon." It is used when the benefit under review is non-compulsory and typically neither provided by the receiving country nor by the employer
- No means "not provided at all" by the receiving country or the employer

Short-Term Benefits

- As most social protection benefits in the USA, short-term benefits are non-compulsory and employer-sponsored, meaning that they are subject to private arrangements agreed in the contract between the employer and his employee without federal obligations for universal coverage. Health care benefits and sick are typically provided by employers
- Short-term benefits in GCC countries are generally employer-sponsored. When this is the case, it is compulsory for the employer to provide basic health coverage for his employee

Work-Related Benefits

- Predeparture training and emergency repatriation costs are typically not covered by receiving countries but increasingly part of bilateral labor agreements or memoranda of understanding as in the seasonal migration schemes in Australia and New Zealand with selected Pacific Islands
- Repatriation in case of death or permanent disability is generally the responsibility of the employer. If the employer does not repatriate the employee or his body, the department in charge returns the employee or his body and recovers the expenses from the employer through administrative means
- There is usually no legal obligation for facilitated access to finance and remittance transfer, but reportedly employers assist their employees in opening bank accounts and finding remittance transfer mechanisms at the lowest costs

countries plus Singapore[a, b]

				Family benefits		Long-term benefits
Wage equality with natives/ vacation pay/ antidiscrimination measures	Unemployment benefits/ work-specific labor market integration/ training	Severance pay/ end-of-service benefits	Provisions in case of bankruptcy/ insolvency fund	Health care benefits (only available in the receiving country when covered by the receiving country)	Family allowance or similar (only available in the receiving country when covered by the receiving country)	Pension benefits/ portability provisions (Yes applies only selectively in case of bilateral/multilateral portability agreements)
YES/yes/yes	No/NC/yes	No/NC	No/no	No	No	Yes/yes
Yes/no/yes	No/NC/yes	No/yes	Yes/no	No	No	Yes/yes
Yes/yes/yes	No/NC/NC	No/no	Yes/yes	No	No	Yes/yes
Yes/no/yes	No/NC/NC	NC/NC	No/no	No	No	No
Yes/yes/yes	Yes/NC/NC	Yes/yes	Yes/yes	Yes	Yes	Yes/yes
Yes/yes/yes	Yes/NC/NC	No/NC	Yes/yes	Yes	Yes	Yes/yes
Yes/NC/yes	No/NC/NC	NC/NC	No/no	NCU	NCU	No
Yes/yes/yes	Yes/NC/yes	Yes/yes	Yes/no	Yes	Yes	Yes/yes
Yes/yes/yes	No/NC/NC	Yes/NC	No/no	No	No	No
Yes/yes/yes	Yes/NC/NC	Yes/yes	Yes/yes	Yes	Yes	Yes/yes
Yes/yes/yes	Yes/NC/NC	No/NC	Yes/yes	Yes	Yes	Yes/yes
Yes/yes/yes	Yes/NC/NC	No/NC	Yes/yes	Yes	Yes	Yes/yes
Yes/NC/yes	No/NC/NC	NC/NC	No/no	NC	NC	Yes/yes
Yes/yes/yes	No/NC/NC	Yes/NC	Yes/no	No	No	Yes/yes
Yes/yes/yes	NC/yes/yes	No/no	Yes/yes	Yes	Yes	Yes/yes
Yes/yes/yes	Yes/NC/NC	Yes/yes	Yes/yes	Yes	Yes	Yes/yes
Yes/yes/yes	No/NC/NC	NC/NC	No/no	Yes	Yes	No`
Yes/yes/yes	No/NC/NC	Yes/NC	No/no	No	No	No
Yes/yes/yes	Yes/NC/NC	No/NC	Yes/yes	Yes	Yes	Yes/yes
Yes/yes/yes	Yes/NC/NC	Yes/NC	Yes/yes	Yes	Yes	Yes/yes
Yes/NC/yes	No/NC/NC	NC/NC	No/no	NC	NC	Yes/yes

- The information provided on severance pay and end-of-service benefits builds on the updated severance pay matrix in Holzmann et al. (2012)
- Provisions in case of bankruptcy mean that pending wages and other worker's entitlements upon redundancy are guaranteed by the receiving countries in case of employer's insolvency, either through a government program such as the Wage Earner Protection Program in Canada or through wage guarantee funds or insolvency funds financed through employers' contribution as in the European countries under review. Such provisions are absent in the USA or New Zealand, where employees' wages and redundancy pay are however considered as preferential claims in bankruptcy procedure

Family Benefits

- When covered by receiving OECD countries, family benefits are only available in the host country and therefore exclude family members left behind in the home country
- Family benefits are generally employer-sponsored in GCC countries except for health care of family members residing in the UAE

Long-Term Benefits

- Portability provisions refer to exportability and totalization of pension claims as included in bilateral or multilateral social security agreements. Yes therefore applies selectively for nationals from countries with which the receiving country under review has signed a social security agreement
- Long-term benefits such as pension benefits in GCCs are typically non-compulsory and uncommon except for Oman, Saudi Arabia, and Bahrain where expatriates are explicitly not covered

Notes

[a]For the full Inventory, see Holzmann and Pouget (2010a)

[b]All the information compiled in this inventory comes from available governmental sources, World Bank and ILO publications (mainly the ILO Migration Survey 2003 and Holzmann et al. (2012) for severance pay and end-of-service benefits) and is subject to revision as legislation evolves. In this sense, the inventory is a living document. The focus is on legal social protection provisions and not on the actual implementation of such provisions

Table 5.6 Summary of inventory of social protection in selected temporary migration schemes – 6 Gulf Cooperation

		Short-term benefits			Work-related benefits			
Skill level	Country	Health care benefits	Work injury benefits/ sick pay	Maternity leave/ paternity leave	Predeparture info and training	Travel provisions/ housing provisions	Emergency repatriation/ cost coverage	Access to finance/ facilitated remittance transfers
Low- to middle-skilled	Kingdom of Bahrain	Yes	Yes/yes	Yes/NC	NCU	Yes/NC	NCU	NCU/NCU
	State of Kuwait	Yes	Yes/yes	Yes/NC	NCU	Yes/yes	NCU	NCU/NCU
	Sultanate of Oman	Yes	Yes/yes	Yes/NC	NCU	Yes/yes	NCU	NCU/NCU
	State of Qatar	Yes	Yes/yes	Yes/NC	NCU	Yes/yes	Yes	NCU/NCU
	Kingdom of Saudi Arabia	Yes	Yes/yes	Yes/yes	NCU	Yes/yes	Yes	NCU/NCU
	United Arab Emirates	Yes	Yes/yes	Yes/no	NCU	Yes/yes	Yes	NCU/NCU
Middle- to high-skilled	Kingdom of Bahrain	Yes	Yes/yes	Yes/NC	NCU	Yes/NC	NCU	NCU/NCU
	State of Kuwait	Yes	Yes/yes	Yes/NC	NCU	Yes/yes	NCU	NCU/NCU
	Sultanate of Oman	Yes	Yes/yes	Yes/NC	NCU	Yes/yes	NCU	NCU/NCU
	State of Qatar	Yes	Yes/yes	Yes/NC	NCU	Yes/yes	Yes	NCU/NCU
	Kingdom of Saudi Arabia	Yes	Yes/yes	Yes/yes	NCU	Yes/yes	Yes	NCU/NCU
	United Arab Emirates	Yes	Yes/yes	Yes/no	NCU	Yes/yes	Yes	NCU/NCU

See General Comments of Table 5.5

Council Countries[a, b]

				Family benefits		Long-term benefits
Wage equality with natives/ vacation pay/ antidiscrimination measures	Unemployment benefits/ work-specific labor market integration/ training	Severance pay/ end-of-service benefits	Provisions in case of bankruptcy/ insolvency fund	Health care benefits (only available in the receiving country when covered by the receiving country)	Family allowance or similar (only available in the receiving country when covered by the receiving country)	Pension benefits/ portability provisions
No/yes/no	Yes/NC/Yes	Yes/yes	No/no	NC	NC	No/NC
No/yes/no	No/NC/NC	YES/YES	No/no	NC	NC	NC/NC
No/yes/no	No/NC/yes	NC/yes	No/no	NC	NC	No/NC
No/yes/no	No/NC/yes	NC/yes	No/no	NC	NC	NC/NC
No/yes/no	No/NC/yes	Yes/yes	No/no	NC	NC	No/NCU
No/yes/no	No/NC/NC	Yes/yes	No/no	NC	NC	NCU/NC
No/yes/no	Yes/NC/yes	Yes/yes	No/no	NC	NC	No/NC
No/yes/no	No/NC/NC	Yes/yes	No/no	NC	NC	NC/NC
No/yes/no	No/NC/yes	NC/yes	No/no	NC	NC	No/NC
No/yes/no	No/NC/yes	NC/yes	No/no	NC	NC	NC/NC
No/yes/no	No/NC/yes	Yes/yes	No/no	NC	NC	No/NCU
No/yes/no	No/NC/NC	Yes/yes	No/no	Yes	NC	NCU/NC

place in underdeveloped financial markets. The impact of the recent financial crisis, concerns with employer compliance and reputation, and a desire to become more attractive to a foreign labor force have motivated GCC countries to think about alternatives and the Emirate of Dubai to commission a think piece for reform (World Bank 2011). Briefly, the proposed three key options for consideration are as follows:

- Transform the (unfunded) EoSB into a funded defined contribution (pension) scheme for all temporary workers. Such a reform should improve compliance by the employer, offer easy portability, and allow for integration with defined contribution/annuity-paying schemes in the country of origin. To achieve a reasonable replacement rate target, the equivalent contribution rate would need to be increased (through first-time contributions by the employee and/or higher contributions by the employer).
- Supplement the (unfunded) EoSB for repeat migrants (e.g., returned, extended, or renewed contracts) with voluntary funded defined contribution schemes in which contributions by the employee are matched by contributions by employers (with a ceiling), with a default option in which the temporary migrant is automatically enrolled with a basic contribution (say 5%) unless he/she opts out. In the development discourse, the potential of matching defined contribution (MDC) schemes to increase benefit coverage and/or benefit adequacy has raised hopes that have been fostered by recent pilots in China and India.[4]
- Use the amounts during accumulation as mobility savings accounts that allow temporary migrant workers to search for a new employer after becoming unemployed/dismissed. The alternative is for the migrant typically to return to his home country and for the employers to seek new migrant workers abroad. Both are linked to high transaction costs.

Under all options, there needs to be the assurance that appropriate financial instruments for retirement savings are available, that financial institutions are well regulated and supervised, that some safeguards for the investors/migrants are available, and that investors are provided with minimum financial capability to make basic choices. Supporting financial sector development has been a mainstay of the World Bank and other development agencies for years. More recently, the development institutions have started to embrace the need for better financial capability and financial education, although the avenues to achieve this are not straightforward (see Holzmann 2010), and there is still an insufficient linkup of these initiatives with migration-related concerns such as the topic of this chapter.

[4] For the presentations of a recent World Bank conference on MDCs, see http://go.worldbank.org/2VD4ZF4MK0 or wait for the forthcoming publication (Holzmann and Hinz 2012, under preparation).

Possible Next Steps

The report commissioned by the Ministry of Labor of the UAE is the first of this kind to approach social protection for temporary migrants against the background of a conceptual framework and a review of a concept- and template-based inventory of social protection provisions and other migration-related services in selected migration-intensive receiving and sending countries. As a result, the approach and conclusions need to be reviewed and discussed in and outside the migration and development community and substantiated by further investigation.

GFMD could play a crucial role in initiating or facilitating work in priority areas, help create support among member countries for institutions that plan to address key issues, and include topics in the agenda of technical meetings for discussion and review.[5] Proposed possible next steps include the following:

- An extension of the inventory to other high-intensity and a number of low-intensity migration-receiving countries with and without special temporary migrant worker programs, to increase the sample size and gain more confidence in the conclusions, or their need for adjustment.
- A full review of legal provisions in the country of origin (such as the welfare funds in South and East Asia) for which the information base has been limited due to issues of access and language.
- A broader international discussion and further theoretical and empirical investigations on the suggested move toward individual accounts for social protection benefits for reasons of coverage and adequacy, incentives, and portability across professions, sectors, and countries (see Holzmann and Koettl 2011). Such an approach could go beyond unemployment and retirement savings accounts and include also health care and other benefits with a major (actual or notional) pre-funding component (Werding and McLennan 2011).
- A progression from the review of legal protection to an evaluation of effective protection. For existing programs, this can be done to a limited extent with corridor case studies between sending and receiving countries, ex post evaluation approaches and preferably with the inclusion of ex ante monitoring and evaluation (M&E) techniques. But to be truly effective, rigorous M&E must be built into the program design. The June 2011 workshop on M&E in Marseille organized by the World Bank jointly with IOM and GFMD is a very promising development and served as a start for an envisaged joint World Bank-IOM work program on the topic.
- The empirical information derived from rigorous M&E would lead to a better understanding of the needs, preferences, and limiting circumstances of social

[5] Such work could be commissioned with a slowly increasing number of policy research centers that focus on migration and social protection issues such as the Labor Mobility Program (by World Bank and IOM at Marseille Center for Mediterranean Integration), IZA (Bonn), Migration Policy Center (University of Florence), and Center on Migration, Policy and Society (University of Oxford).

protection programs for temporary migrants. This in turn would allow for better tailoring of the provisions and services for this increasingly important workforce across the world.

- Separate and joint pilots in sending and receiving countries that attempt a better coordination of social protection coverage for temporary migrants with regard to key benefits that have proven difficult to establish (such as for health and pensions).

To guide any further discussion, analysis, and next steps, the following questions may need to be considered:

1. How can one best accommodate the differences and similarities in risk profiles and social protection needs across the various types of migrants and social protection benefits' supply in sending and receiving countries for their workforce?
2. What is the role of sending countries in the social protection of their temporary migrants abroad in view of the potential trade-offs between the scope of social protection in the receiving countries and the size of remittances?
3. What is the role of an initially restricted access to social protection by temporary migrants as a signaling and screening device, and what are the welfare implications and trade-offs for sending and receiving country and the migrants?

References

Abella, M. (2006, June 28–30). *Policies and best practices for management of temporary migration.* Paper prepared for the International Symposium on International Migration and Development, UN Population Division, Department of Economic and Social Affairs, Turin, Italy.

Agunias, D., & Newland, K. (2007). *Circular migration and development: Trends, policy routes, and ways forward.* Washington, DC: Migration Information Source, MPI (Migration Policy Institute).

Agunias, D. R., & Ruiz, N. (2007). *Protecting overseas workers: Lessons and cautions from the Philippines*. Washington, DC: Migration Policy Institute.

Avato, J., Koettl, J., & Sabates-Wheeler, R. (2009). *Definitions, good practices, and global estimates on the status of social protection for international migrants* (WB Social Protection Discussion Paper No. 0909, Social Protection Unit). Washington, DC: World Bank.

Del Rosario, T. (2008). *Best practices in social insurance for migrant workers: The case of Sri Lanka* (ILO Asian Regional Programme on Governance of Labour Migration Working Paper 12). Geneva: ILO.

Forteza, A. (2010). The portability of pension rights: General principles and the Caribbean case. *Development Policy Review, 28*(2), 237–255.

Holzmann, R. (2010, August). *Bringing financial literacy and education to low and middle income countries: The need to review, adjust, and extend current wisdom* (IZA Discussion Paper No. 5114). Retrieved from http://ftp.iza.org/dp5114.pdf

Holzmann, R., & Hinz, R. (Eds.). (2012). *Matching defined contribution (MDC) Schemes: A promising pension design to enhance converge and benefit level?* Washington, DC: World Bank (under preparation).

Holzmann, R., & Jørgensen, S. (2000). *Social risk management: A new conceptual framework for social protection, and beyond* (WB Social Protection Discussion Paper 0006, Social Protection Unit). Washington, DC: World Bank.

Holzmann, R., & Koettl, J. (2011, May). *Portability of pension, health, and other social benefits: Facts, concepts, issues* (World Bank, Social Protection Discussion Paper No. 1110). Washington, DC: World Bank.

Holzmann, R., & Pouget, Y. (2010a, November 8–11). *Social protection for temporary migrant workers: Conceptual framework, country inventory, assessment and guidance.* A Study prepared for the Global Forum of Migration and Development (GFMD) Working Group on Protecting and Empowering Migrants for Development, in support of GFMD Puerto Vallarta, Mexico.

Holzmann, R., & Pouget, Y. (2010b, December). Towards an objective-driven system of smart labor migration management *Economic Premise 42*, Washington, DC: World Bank.

Holzmann, R., Koettl, J., & Chernetsky, T. (2005). *Portability regimes of pension and health care benefits for international migrants: An analysis of issues and good practices* (WB Social Protection Discussion Paper No 0519, Social Protection Unit). Washington, DC: World Bank.

Holzmann, R., Pouget, Y., Vodopivec, M., & Weber, M. (2012, June). *Severance Pay Programs around the world: Rationale, status, and reforms* (World Bank, Social Protection Discussion Paper No. 1111). Washington, DC: World Bank.

ILO. (2006). *ILO multilateral framework on labour migration: Non-binding principles and guidelines for a right-based approach to labour migration.* Geneva: ILO.

Koettl, J. (2006, May). *The relative merits of skilled and unskilled migration, temporary and permanent labor migration, and portability of social security benefits* (World Bank, Social Protection Discussion Paper, No 0614).Washington, DC: World Bank.

Luthria, M. (2011, January). Labor mobility for the poor: is it really possible? *Economic Premise No. 45.* Washington, DC: World Bank.

Martin, P. (2003). *Managing labor migration: Temporary worker programs for the 21st Century.* Geneva: (ILO Institute for Labour Studies).

Martin, P. (2005). *Towards effective temporary worker programs: Issues and challenges in industrial countries* (International Migration Paper No 89, International Migration Program). Geneva: ILO.

OSCE (Organization for Security and Co-operation in Europe)., IOM (International Organization for Migration)., & ILO (International Labour Office). (2007). *Handbook on establishing effective labour migration policies: Mediterranean edition.* Vienna/Geneva: OSCE/IOM and ILO.

Ruhs, M. (2002). *Temporary foreign worker programmes: Policies, adverse consequences, and the need to make them work* (Perspectives on Labour Migration No 6, Social Protection Sector, International Migration Program). Geneva: ILO.

Ruhs, M. (2005). *The potential of temporary migration programmes in future international migration policy.* Paper prepared for the Policy Analysis and Research Programme of the GCIM (Global Commission on International Migration), Geneva.

Ruhs, M. (2011). *Openness, skills and rights: An empirical analysis of labour immigration programmes in 46 high- and middle- income countries.* Centre on Migration, Policy and Society (COMPAS), University of Oxford, WP-11–88.

Ruiz, N., & Agunias, D. R. (2008). *Protecting temporary workers: Migrant welfare funds from developing countries* (Migration and Development Brief No. 7, Migration and Remittances Team). Washington, DC: The World Bank.

Sabates-Wheeler, R. (2009). *Extending social security coverage: Good practices, lessons learnt and ways forward* (Working Paper No. 12). Geneva: ISSA.

Vonk, G., & Kapuy, K. (2008). Three approaches to refunding social insurance contributions to temporary migrants workers: Is there an attractive policy alternative? *European Journal of Social Security, 10*(3), 227–247.

Werding, M., & McLennan, S. (2011, August). *International portability of health-care benefits: Concepts and experience* (World Bank, Social Protection Discussion Paper No. 11XX). Washington, DC: World Bank.

World Bank. (2001). *Social protection sector strategy: From safety net to springboard.* Washington, DC: World Bank.

World Bank. (2011, June). *Social protection for expatriate workers in Dubai: Retirement and mobility savings accounts.* Report prepared for the Emirate of Dubai. Washington, DC. Confidential.

Chapter 6
Textbox 3: Strengthening Migration Health Management in Sri Lanka

IOM Geneva* and IOM Sri Lanka

"Migration is important in relation to Sri Lanka's vision of becoming a vital hub in the global and regional economy articulated in the catchy and emotive idea of Sri Lanka turning into the 'Wonder of Asia'. Despite the many positive effects of migration on our society and economy we must ensure balance of negative impacts of migration with benefits. This will depend on protecting the health and wellbeing of migrant populations their families and the host populations". (Maithripala Sirisena, Minister of Health, Sri Lanka.)

Notwithstanding the absence of costed health care models for the many types of migration today, there are strong arguments for granting migrants access to existing public health care services. These include cost-effectiveness, human rights, social integration, public health, and commitment to international resolutions and declarations.[1] Countries like Mexico and Sri Lanka are ensuring better access to health care services for their own nationals while abroad, while many other countries hosting migrants – including middle-income countries such as Thailand and Argentina – have developed schemes to give nonnationals in their countries better access to health care services.[2]

In Sri Lanka, the government authorities have increasingly become aware of the major migration health challenges and are committed to meeting the health demands of the various types of migrant and mobile populations.

---

*Source: IOM Geneva and IOM Sri Lanka, with input and endorsement from the Ministry of Health of Sri Lanka.

[1] Including 61st World Health Assembly Resolution (WHA 61.17, 2008), which calls upon member states to promote the "health of migrants."

[2] See Annex 1 of RT 2.1 GFMD.

IOM Geneva (✉) • IOM Sri Lanka,
Ministry of Health, Sri Lanka

I. Omelaniuk (ed.), *Global Perspectives on Migration and Development: GFMD Puerto Vallarta and Beyond*, Global Migration Issues 1, DOI 10.1007/978-94-007-4110-2_6,

The government recognizes the potential health impacts of the full range of migrants and mobile populations. These stem from three flows of migrant populations (inbound, outbound, and internal) as well as impacting the families "left behind."

An estimated 1.8 million Sri Lankans are temporarily employed abroad,[3] approximately 18% of the total labor force. The foreign employment industry continued to be the second largest earner of foreign exchanges.[4] In addition, Sri Lanka hosts a large number of foreign migrant workers, internal migrants, and displaced persons as a result of years of civil strife, returning refugees and failed asylum seekers, international students, and health professionals seeking to work abroad.

Currently, the direct health costs related to the various types of migration during the different stages of the migration process are borne by different actors including migrants themselves, the government of Sri Lanka, the government of the destination country, and employers in the destination country. There is currently no comprehensive picture of the costs involved and how these costs are distributed among the different actors.

The government is concluding a national study on migration health (August 2011), which reviews the health status and determinants on outbound migration, families left behind, and inbound and internal migration. These studies will provide empirical evidence on the health and social impacts on migration health and development. Studies on costs to health systems and financing of risks and factors stemming from migration health are planned as the second phase of the research agenda.

What makes Sri Lanka an interesting case study is that it has acknowledged that migration health cuts across many different sectors and actors and that vertical and horizontal approaches are needed to ensure that migration does not negatively affect development of migrants, their families left behind, and source and destination countries. In line with the 2008 WHA Resolution "the health of migrants," the Ministry of Health is in the process of developing a *national policy for migration health* and has established an interministerial coordinating framework on migration health engaging the key government ministries involved at the interface of migration, health, and development (see diagram below). Sri Lanka is also the first country to present the "report card" on the progress made on the resolution of the health of migrants at the World Health Assembly in 2011.

[3] Central Bank of Sri Lanka, Annual Report 2008.

[4] SLBFE Annual Statistical Report 2009.

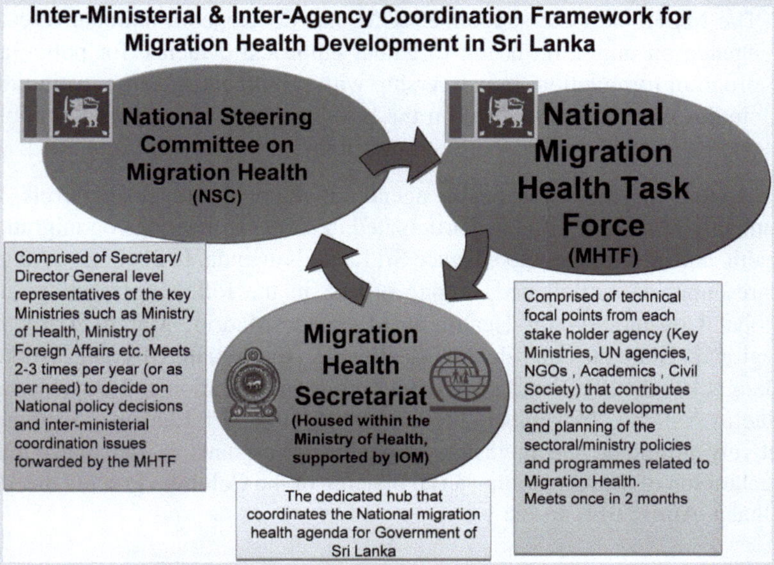

The government of Sri Lanka has undertaken a mapping of relevant ministries and of domestic legislation and health, legal, and social protection services provided to different types of migrants. The MOH has a strong coordinating role, but migration health is a shared responsibility: for example, for outbound migration, a key ministry involved is the Ministry of Foreign Employment Promotion and Welfare, and the Sri Lanka Bureau of Foreign Employment is involved in migrant worker welfare and health insurance schemes for all Sri Lankan migrants working abroad.

Recognizing that migration health has multi-sectoral determinants and impacts, an interministerial *"whole of government approach"* involving 12 key government ministries was adopted to advance Sri Lanka's Migration Health Agenda at the end of 2009. Remarkable progress has been made since then. Key achievements to date include:

- The Ministry of Health with other government ministries and IOM developed a national reintegration health strategy, including practical health assessment protocols for approximately 87,000 Sri Lankan refugees returning from southern India. The returnees are linked to primary health care services upon return to Sri Lanka and are also provided with health information (e.g., a booklet specially produced by MOH and IOM) prior to departure.
- The Ministry of Health and the immigration controller general are jointly developing technical guidelines and protocols for *"health requirements for long stay visa applicants"* and establishing a visa health unit for long-stay resident visa holders entering Sri Lanka.

- The MOH, with the support of IOM, has developed a national research agenda on migration health to gather empirical evidence for policy and program formulation. In partnership with prominent research institutions, studies were commissioned on the health impacts of outbound, inbound, and internal migration, to be completed in 2011 and released in 2012.

Moving the migration health agenda at regional and global levels, Sri Lankan government officials participated in strategic meetings on migration, health, and development to advance Sri Lanka's agenda. In 2010, government representatives participated, among others, in the following meetings: the Global Consultation on Health of Migrants (Madrid, March 2010), the Regional Dialogue on Health Challenges for Asian Migrant Workers (Bangkok, July 2010), and the Global Forum on Migration and Development (Puerto Vallarta, November 2010). In 2011, the Sri Lankan government actively promoted the inclusion of health of migrants in the final Dhaka Declaration of the recent ministerial meeting of the Colombo Process meeting (Dhaka, April 2011, article 1.8).

(Contact for more information on Migration Health Programs in Sri Lanka: Kolitha Wickramage, IOM Sri Lanka. E-mail: kwickramage@iom.int)

Chapter 7
Migration, Gender, and Family

Juan Carlos Calleros Alarcon*

Introduction

Despite growing evidence about the importance of family and gender for migration and development, there is little international discussion about how these four fields concretely interface with each other. Migration studies and policies traditionally have focused on individuals rather than on the family as a unit of analysis or recipient of migration services and benefits. For example, there is little discussion about the social and family costs of migration, considering that family ties are the main impetus for migrants to send remittances to the communities of origin.

Without replacing the current focus on the migrant per se, the participants of GFMD 2010 were unanimous in recognizing the need to bring a new analytical perspective to this field based on family and gender. The themes were welcomed as a new way of looking at the critical links between migration and development. Since family welfare and gender mainstreaming are pivotal to achieving the Millennium Development Goals (MDGs), this is in line with more recent UN efforts to redefine development from the more human and social angle.

The reason why family and gender have remained unlinked in migration policies and in other fields is their different ideological origins. Family has been advanced by followers of a more conservative approach, while gender equality and the gender perspective have clearly been an important concern of progressive thinkers. It is not the first time that such an apparent dichotomy has appeared on issues close to both

*This chapter was prepared by Dr. Juan Carlos Calleros Alarcon, Researcher at the Center for Migration Studies (Investigador del Centro de Estudios Migratorios), Mexico City. The author acknowledges Salvador Berumen, Gloria Moreno-Fontes, Blandine Mollard, and Helen Schwenken for their reviews of the first version of this chapter.

J.C.C. Alarcon (✉)
Center for Migration Studies, INM, Mexico City, Mexico
e-mail: jccalleros@inami.gob.mx

I. Omelaniuk (ed.), *Global Perspectives on Migration and Development: GFMD Puerto Vallarta and Beyond*, Global Migration Issues 1, DOI 10.1007/978-94-007-4110-2_7,

approaches, for example, when in 1994, the Argentine Mothers of the May Square found it odd to sit on the same side as the feminists during the Cairo UN Summit on Population and Development. This uneasy feeling arose because the Argentine Mothers harbored conservative points of view about gender issues, even if they were radical activists on the "disappearance" of their sons and daughters.[1]

In a similar fashion, some may find it difficult to create synergies when a theme with a conservative basis, such as family, and an upfront theme, such as gender equality, come together on migration issues. Also, the intersection of both perspectives with migration and development is too recent a conceptual path to achieve clear, comfortable synergies all around.

Since members of the Forum decided to keep the themes together, rather than have one round of discussions on gender and migration, and another on family and migration, the background paper and final debate of the GFMD Roundtable (RT) on "Family, Gender and Migration" in Puerto Vallarta took a dialectic approach to finding conceptual synergies among them. They looked for a synthesized framework, to analyze the migration and development connection through the lenses of family and gender together. Such a solution has the added advantage of permitting a deeper analysis of migrant women, both as part of the family (as spouses, daughters, mothers, or companions) and as migrants traveling on their own. Critical aspects of female disadvantages within the family – in the country of origin (COO) and the country of destination (COD) – and of the specific place of gender in migration (e.g., women staying behind or migrant women sending remittances) could be uncovered by this approach.

This new perspective takes account of the so-called "mezzo" level of analysis – between the individual and the social or economic structures – and recognizes the transformations and fragmentation that family structures undergo as a result of migration. Today, transnational families – with one or several members living in different countries – are a common feature of complex global labor market developments (e.g., Turks in Germany, Moroccans in Spain, Mexicans in the United States). They are also often a consequence of, and can be affected by, restrictive migration and residency policies of countries of destination. Mixed-status families in the same location are one of the new migration realities that challenge the traditional concept of family and policies tailored to this.

A major concern behind the theme of Family, Gender, and Migration – which also arose during the August 2010 session of the Regional Conference on Migration (RCM) – was that policies are not keeping pace with new migration features, and this can have serious development impacts. In particular, new policies and practices are needed to preserve the principle of family unity, a fundamental right under international law, in light of these new trends. Children and adolescents traveling on their own are also a missed target for migration policies. The feminization of

[1] See Fisher (1989), p. x: "(…) what is normally considered a conservative force in society (the mothers), was converted into a force of opposition to military rule (…) Their participation with the Mothers forced them to modify their traditional role as mothers and changed their perceptions of their roles within their family and society."

migration and the "Penelope" syndrome of women staying behind[2] have been widely recognized, but few gender-sensitive actions have been taken within existing policies to address these trends. All of these apply equally to countries of origin, transit, and destination.

Against these trends, and the need for concerted actions and partnerships with international organizations, civil society, and the private sector throughout the migration cycle, this chapter draws on the issues and outcomes of the Background Paper and discussions of Roundtable (RT) 2.2 in Puerto Vallarta in 2010. It also reflects on the civil society papers and preliminary discussions in Mexico City at the Bancomer Center in July 2010. The chapter is divided into five segments dealing with (a) the transnational family, (b) unaccompanied migrant children, (c) the gender perspective, migration, and development, (d) how to test the main assumptions, and (e) recommendations and how to factor them into coherent policy making.

The Transnational Family

Migration patterns have shown that families can find themselves in three transnational situations: (a) when members of the family are spread across countries, (b) when both parents are abroad while children remain in the country of origin, or (c) one parent remains at home while the other works abroad. It was generally recognized in Puerto Vallarta that the transnational circumstances of these families can have deep, long-term social and economic impacts, particularly on the education and health of children, the well-being of spouses left behind, and the capacity of the household to continue existing as a relevant social unit for the community, either in the country of origin or destination.

Since transnational family members reside in different countries, even in different continents and environments, the meaning of "family" and the social functions within the family unit are challenged and redefined. The very definition of family is culturally differentiated and cannot be taken for granted today, since new forms of social ties and coexistence have evolved that could be considered "family relations," even if different from the traditional concept of the "nuclear family."

The discussion about any redefined meaning of "family" could be endless. Even the definition in the International Convention on the Protection of Rights for All Migrant Workers and Members of their Families (CMWF) is not cross-cultural or universal since it depends on applicable legislation or agreements between the countries concerned.[3] For this reason, the RT 2.2 Background Paper avoided any

[2] The Penelope Syndrome can lead to somatic and psychic disorders among spouses left behind that affect the family and the community (Castro 2010).

[3] Article 4 of the CMWF states: "For the purposes of the present Convention the term 'members of the family' refers to persons married to migrant workers or having with them a relationship that, according to applicable law, produces effects equivalent to marriage, as well as their dependent children and other dependent persons who are recognized as members of the family by applicable legislation or applicable bilateral or multilateral agreements between the States concerned" (underlined by the author) (United Nation General Assembly 1990).

reference to a specific family definition. Instead, the concept of "family structure" is preferred, where persons are part of, and support and rely on, a family structure across countries, whatever the specific definition of "family" applicable to them. This implies that access of migrants to their family structure, in the COO and the COD, should be the target for any family-sensitive migration policies. This may start with the nuclear family, but it could go beyond that, depending on the applicable legislation and the cultural implications for family units.

Applying the concept of "family structure," the phenomenon of mixed-status families living in the same household is often a reflection of the country of destination's residency laws. These can have significant consequences for the family unit, for example, even if a child of immigrants is a national by birth in the destination country, he/she may be deprived of health and education rights if the parents are in irregular migration status. Raids and deportations also have more tragic consequences for migrant families with mixed migration status. This is far more significant than just the removal of irregular migrants and the enforcement of migration law. It affects the family unit profoundly through forced separations and consequent disruptions to the family life.

Traditionally, migration policies have considered families as an adjunct to the migrants, and migration systems are designed to issue visas or work permits to a main migrant who may or may not have a nuclear family. If he or she does have a family, it is of secondary importance since the visa or work permit is strictly *ad personum*. The principal migrant may want to bring his or her family along, but is usually discouraged from doing so unless visas are issued to all accompanying family members, and proper accommodation arrangements have been made for this purpose. For some migrants, it could simply be beyond their reach to have their families with them, either because of the difficulty of obtaining visas and/or work permits for them or on financial grounds, at least until the first salaries are received and savings commence.

Transnational living is a common condition for many families around the world. It is a strategy for family survival based on temporary and circular mobility of family members across borders. This transnationalism is essential for remittances. It also stimulates growth of services, communication, and transportation across borders and creates favorable conditions for nostalgic trade, that is, imports of food, beverages, clothes and fabrics, handicrafts, or other goods produced in the country of origin. Nostalgic trade could be, and is usually accompanied by, the export of instruments, machinery, toys, books, and technology from the destination country, starting a trade currency based precisely on transnational communities (Ariza and Portes 2010, p. 35).

In the meantime, migrants are forced to be separated from their family for long periods, often many years, while the family left behind is deprived of the presence of the main provider (male of female), and the migrant is deprived of the emotional, social, and psychological support that derives from cohabitation with the family. The Mexican President, Felipe Calderón, described this problem in his speech at the Puerto Vallarta meeting:

> (We) know that for every migrant leaving, it means a father who will not see his kids, or kids that will not see their father again; because we know that for every migrant leaving, a

> prospective community leader is leaving; because we know that the hard workers, the dearest, the bravest are leaving; because we are sad to see our people and communities transformed, practically, in deserted towns where there are only old persons and some women.
>
> Migration hurts us. And we know that migrants take the ugly decision to leave their soil and their family, sometimes forever, because they do not have the opportunity to give something to their offspring.

These are some of the social costs of migration, and the emotional aspects of families torn apart by migration when the father leaves; but they also apply when the mother is the principal migrant. Regarding the connection to development, the assumption in Puerto Vallarta was that migrant workers who have access to their usual family structures are potentially more productive, healthier, and better adjusted than those isolated from their family. It follows, then, that fostering the well-being of migrants within a family setting, either through family reunification, circularity, or closer cross-border communication, can have positive externalities for all, such as reduced stress for the migrants and more regular availability of migrant workers for employers in the country of destination.

For families staying behind, migration is usually a defining experience. As a result of emigration of the head of the family, wives who stay behind can be exposed to a number of social stress factors, for example, by assuming full responsibility for the education of the children and other multiple household chores, while fulfilling roles usually reserved for their husbands – political representation or certain godfather duties – or dealing with pressures from members of the extended family about how to spend or invest remittances. Such pressures can have physical and psychological consequences that in certain cases should be addressed by health providers. Evidence is mounting about the "Penelope" syndrome, where women who stay behind are subject to physical and emotional stress and may suffer somatic and psychic disorders in communities with high rates of male migration.

In some contexts, the migration of men results in the family moving in with the husband's parents, resulting in reduced freedom of movement for the women and a lack of control over spending and investment of remittances by the woman. Another outcome related to this is the sudden loss of alternative sources of income for the women left behind, followed by extreme dependency on the remittances.[4]

In many cases, migrants initially perceive the separation of their family as a temporary phenomenon, where transnational ties are only necessary for the time required to save some money before returning home to the families. However, it is common for migrants to prolong their stay abroad because the precariousness of their income, in view of high living costs, retards their saving capacity. Restrictive immigration policies are another reason for extending their stay, sometimes over many years. For undocumented migrants in particular, these lower their chances to reenter their host country if they go home, and so they stay on, often indefinitely. Consequently, transnational families remain separated for longer periods than expected.

---

[4] See "Abandoned wives of Tajik labor migrants," available at: http://www.iom.tj/pubs/abandoned_wives_English.pdf, and Boehm (2008, p. 98–107).

When it is clear to transnational families that the separation can be much longer than planned, reunification is the main option to be followed. In such cases, part or the rest of the family may join the main migrant – in authorized or unauthorized ways – with the intention of either settling or staying indefinitely in the country of destination. Over time, some members of the family may enjoy legal status while others may have no authorized status. Children develop social networks of their own, and the main sociocultural references, such as language, social interaction, celebrations, and identity, increasingly become linked to the host country, while the first generation of migrants usually maintain their social and cultural ties with the place of origin.

This scenario is reflected poignantly in the Mexican documentary film *"Los que se quedan*" ("Those who stay behind"), presented at the GFMD 2010. It depicts a handful of rural Mexican families affected by the out-migration of one or several of their members to "the North," that is, the United States.

Box 7.1 "Los que se quedan"[5]

In the documentary "*Los que se quedan*," one of the Mexicans who left the country explained as his motive that only those who migrated to the North – either regularly or irregularly – ended up owning land or property of their own. Another family explained that the father of three girls wanted to migrate in order to send remittances for the girls' education in a private secondary school. Women and families left behind are shown preparing for the arrival of their spouses after several months of absence.

One of the returning men explains his reluctance to work in Mexico: the small pay is not worth the effort – he was accustomed to being paid ten times more for the same job in the United States. Other families show the fruits of their remittances: bigger and better houses than their neighbors in the same community. But there are also empty houses due to the prolonged absence of the husband and his insistence on regrouping the family in the United States – usually in an irregular manner.

The film shows the transnational nature of rural Mexico, where migration is an effective strategy for social mobility, in terms of land acquisition, education for children left behind, or home repairs and extensions. There are no examples of investment of remittances in local business creation in the film, but only the sadness and stoic sentiment of families who stay behind. The message is clear in the end: they would rather be together as families – in Mexico or in the United States – if the option was available. Another important feature of transnational migration shown in the documentary is its circularity. Traditionally, Mexican migration to the United States has been circular, but today it is more clandestine, with longer periods elapsing for family reunification to happen.

[5] The documentary was financed by Fundación Bancomer-Mexico and directed by Juan Carlos Rulfo and Carlos Hagerman. For the trailer of the film and basic information about it, see: http://losquesequedan.com/

Examples like these clearly show the need for policies to specifically address transnational families within the migration cycle. Migration policies are traditionally designed for individual migrants, especially where they have undocumented status. And even if most developed countries espouse family reunification as an important principle when dealing with documented migrants, many policies still require the main migrant to have been legally resident for 12–18 months before being eligible to apply for family reunification (e.g., Germany and the European Union Blue Card scheme). Additional socioeconomic criteria like a high-income level can also be difficult to meet.

Migrant families have played a secondary role in most migration policies worldwide, but the increasing prominence of transnational families in certain binational migration systems (Mexico-US, Turkey-Germany, Morocco-Spain, Philippines-Gulf States, etc.) advances the case for a more comprehensive family-oriented approach in migration management worldwide. Too few countries consider the needs of families, spouses, and children, either staying behind or accompanying the main migrant, in their policies. Many migration policies are even aimed at the forced separation of parents and children, for example, through raids and expedited deportation of the undocumented parents.

There is a need for more sensible migration policies that do not deport irregular migrants with children who are nationals of the country of destination by birth and that allow for regularization options or special consideration of such cases.[6] For example, the Argentine Migration Law (May 2010, Art. 70) and the new Mexican Migration Law (May 2011, Art. 133) provide permanent regularization for all irregular migrants with Argentine/Mexican partners or Argentine/Mexican offspring.

As discussed in Puerto Vallarta, certain sets of policies can contribute to the well-being of migrants and their families, both in the country of origin and destination. These policies range from migration management to health and social security, education, and employment opportunities. Family friendly policies and measures can also be smart for development since migrant workers who have access to their family structure can be more productive, potentially healthier, and better adapted than those who are alone. Such migrants can be a more certain and predictable workforce from abroad, critical for regional development planning. Hence, some partnerships in CODs offer young people education and employment opportunities at home to reduce the urgency to migrate, while others focus on strengthening family unity across borders.

Among the most outstanding good practices discussed in Puerto Vallarta, concerning partnerships created to address this kind of issue, were the following:

- Portugal recently established an Office for the Support of Family Reunification (GARF) in each of its National Centers for Immigrant Support, in Lisbon, Porto, and Faro. In 2009, these offices assisted 10,831 family reunification migrants, an increase of 17% over 2008.

[6] It only applies to countries where citizenship is obtained on the *ius soli* principle.

- In 2008, Mexico introduced a work permit card for Guatemalan frontier workers (*Forma Migratoria de Trabajadores Fronterizos*). This document allows the migrants to work in any economic sector of the Southern border states, multiple entries, accompanying spouses and/or children to obtain their own card (preserving the unity of working migrant families), family reunification, and circularity of documented migrants. A complementary card for local visitors (*Forma Migratoria de Visitante Local*), also launched in 2008, allows visits of spouses and relatives of Guatemalan workers at all work sites, especially coffee plantations such as in Soconusco, Chiapas.[7]
- The Philippines has established family counseling and welfare services for overseas Filipino workers (OFWs) and their families abroad and at home and organizes OFW Family Circles as part of the psychosocial program for reintegration. To better link emigration and reintegration planning with development, a digital registry of the poorest households with OFWs is being constructed, to inform future comprehensive social welfare services for the families of OFWs.

  Since 2008, the Tulay Project (Tulay means "bridge"), in partnership with Microsoft Co., provides IT training and access to technology for regular communication through Internet applications between overseas workers and their families back home. Nineteen national centers in the Philippines and six centers abroad are already functioning as part of this initiative. During the session in Puerto Vallarta, it was reported that 43,000 Filipinos have been trained under this program.
- In Junín region in Peru, the International Fund for Agricultural Development (IFAD) is funding a pilot program on entrepreneurial leadership aimed at transnational migrant family networks in the region. The objective is to prepare business plans for productive investment of remittances for 80 families in partnership with 10 local governments and 10 Peruvian migrant organizations in the United States and Canada. Twenty business plans for transnational families are in progress, while 13 families have already developed a business plan, including some involving transnational trade with potential trickle-down effects for the communities of origin.[8]
- In partnership with the European Union (EU), Portugal, and Spain, Cape Verde has recently started up a Centre for Migrant Support in the Origin Country (CAMPO), designed to assist Cape Verdean migrants at all stages of the migration cycle, including emigration into the EU, reintegration in the country of origin labor market, and support for migrant-led entrepreneurship.

Most of the discussion in Puerto Vallarta was directed at including family friendly actions, such as the ones described above, in migration policies. One common concern was the need to redefine "family" since it can be nuclear, extended, or involve quite a different membership depending on the country and culture. Even if the

---

[7] These actions are complemented with visits to coffee plantations by labor and migration officials in order to enforce the respect of human and labor rights for Guatemalan workers.

[8] http://www.ifad.org/remittances/projects/lac/peru.htm, consulted on May 30, 2011.

theme was not exhausted in 2010, migrants' access to their customary family structure, or to close relatives, is an important factor in maximizing productivity, physical and psychological health, and adaptability of migrant workers to their host communities. Family reunification should thus be a general orientation of migration policies. Avoiding unnecessary delays in migration processing and halting raids for detention and expedited repatriation of undocumented migrants with family members, especially children, in the country of destination were high among the recommendations in Puerto Vallarta.

Unaccompanied Migrant Children: New Phenomenon, Old Vices

One of the most striking aspects of contemporary international migration is the phenomenon of unaccompanied migrant children. This category mostly covers teenagers between 12 and 18 years of age, who are able to undertake the trip but are highly vulnerable due to their physical limitations and lack of means and knowledge necessary to defend their legal rights. But there are also serious cases of unaccompanied children under 12 years, who are even more exposed to multiple risks and abuses.

Immigration processes, either authorized or unauthorized, increasingly involve the presence of children and adolescents as one of the most vulnerable groups. Yet there is a serious dearth of information about these groups to guide policies and a corresponding lack of policy consideration of these issues. Their plight brings to light the fact that poor or nonexistent social policies in countries of origin are one of the biggest causes of irregular migration, including trafficking in persons, which can undercut the positive effects of migration for development. The lack of such policies is clearly a broader development issue with migration consequences, but more specifically it is a hindrance to the full human development of women, children, and the elderly. New comprehensive policy approaches are needed to address these kinds of interlinked migration/development challenges affecting the most vulnerable at home and abroad.

During the inaugural ceremony of the Puerto Vallarta GFMD, the First Lady of Mexico, Mrs. Margarita Zavala urged participants "to see migration through the eyes of children" since separation hurts children and is a common tragedy for transnational families.[9] The phrase was widely cited during the Civil Society Days and the Governmental Roundtables, and sympathy was widely expressed for this vulnerable group of migrants, who number in the tens of thousands in some migration corridors, such as between Mexico/Central America and the United States.

[9] Another participant in RT 2.2 of the GFMD in Puerto Vallarta urged "to see migration through the eyes of women migrants." In this paper, migration is seen dialectically through the lenses of family (children and women migrant included) and gender perspectives, as explained before.

From the migration management perspective, there are four different groups of minors: (a) those left behind, (b) adolescents traveling alone, (c) the second generation of migrant children in countries of destination, and (d) children traveling with friends or relatives other than their parents or guardians, smugglers, or traffickers (who account for a worrisome number of "unaccompanied children"). In Puerto Vallarta, governments discussed the fact that the lack of regular avenues for migration may push children to migrate alone and expose them to criminal activities, including smuggling of drugs and persons and trafficking in persons.

The case of minors traveling with their guardians or intermediates can take different forms: the family travels within the home country toward another country or border, the minors are victims of trafficking at the hands of their own relatives, or they may have crossed the border without authorization and subsequently been captured and repatriated along with their parents. In the last scenario, undocumented minors can also be traveling alone, but with undocumented parents who are detained in a country of transit for repatriation purposes. In such cases, the detention and attention provided to the minors need to be carried out or supervised by trained personnel, familiar with the specific aspects of child migration and child detention.

The most complex cases are unaccompanied children under 12 years of age, traveling without their guardians or parents precisely because they want to reunite with their families. Most cases are heartbreaking, worthy of inclusion in novels such as Charles Dickens' *Oliver Twist* or Jules Verne's *Petit Bonhomme*. In these fictions, however, there are happy endings after many unfortunate incidents for the two main characters. By contrast, the majority of unaccompanied children in the real world are not always so fortunate, often ending up in painful repatriations, prolonged detention,[10] and, in some cases, child prostitution and trafficking for the purposes of sexual or labor exploitation.

For many, there is no happy ending, and taking care of these little migrants expediently and efficiently is a challenge to be taken up by all affected states and societies as soon as possible. Stronger, more realistic policies are needed specifically for this vulnerable group. Partnerships and joint projects are needed to translate laws into protection systems and policies for complying with the children's best interest in countries of origin, transit, and destination. The GFMD last year called on governments to go beyond just signing treaties and conventions for the protection of children's best interests and vigorously redress inadequate systems of implementation at national level.

For example, despite the widely ratified Convention on the Rights of the Child of 1989, the right to identity for children still poses a major challenge: some 50 million births go unregistered each year (either because of irregular status, fear of detection, or other obstacles, including language barriers). Migrant children are among the

[10] For contemporary histories of children traveling on their own from Central America to the United States, passing through Mexico, see the trailer and the extra footage of the film by Rebecca Cammisa, *Which Way Home*, special jury prize for human rights Michigan 2010, and Academy award nominee 2011, at http://whichwayhome.net/

most severely affected since they can be prevented from accessing health care and education and can be denied essential immunization and health protection among other curtailments of rights.

Also, despite the widely ratified ILO Convention on the Minimum Age of Employment (138) and the Worst Forms of Child Labor Convention (182), child labor is still common in many countries. Observers usually ask why there is such a gap between the international principles and national immigration policies and laws, especially in the case of migrant children. The serious dearth of data on children, particularly undocumented or children of undocumented migrants, was noted in Puerto Vallarta. Only with such data would it be possible to assess if the under-registration of migrant children is due to a gap in resources, knowledge, or commitment.

It was thus agreed in 2010 to ratify all relevant conventions and ensure national implementation and to produce better registries of migrant children in all countries attending the GFMD 2010. Following are some good practices for support and protection of migrant children and adolescents, as reflected in the background documentation and the Roundtable discussion:

- Armenia's Diaspora Youth Mobilization program ("Ari-tun-come home"), which aims to keep young Armenians informed about their homeland and nurture their identity by accommodating and hosting them in Armenian families.
- The Honduran Government's work with a consortium of international organizations (UNDP/FAO/ILO/UNODC/UNFPA/UNICEF/IOM), as well as local youth groups, families, communities, and migrant associations abroad to generate jobs and business opportunities at the local level for young people between 15 and 29 years of age. The aim is to empower youths in countries of origin and reduce their susceptibility to irregular migration.[11]
- The shelters and centers of the National System for Family Development (DIF) since 2007 in Mexico that assist an average 19,500 unaccompanied migrant children and adolescents repatriated from the United States. In addition, a Trust Fund with HSBC Bank, ONG "Save the Children," and DIF has provided scholarships and food support for repatriated Mexican migrant children and resources for migrant transportation, among other actions.
- Mexico's recently established corps of Officers for the Protection of Infants (OPIs) in the National Immigration Office (INM) provides support and services to unaccompanied migrant children and offers training to migration officials from other countries dealing with the protection of children's rights. As at July 2011, some 320 OPIs had provided specialized care to 60,000 unaccompanied migrant children since 2008. The model has been shared with Central American countries. In the past 2 years, the corps has trained OPIs from Guatemala, El Salvador, Honduras, and Dominican Republic, and other Central American countries have requested such training.[12]

---

[11] See the program "Human Development for Youth: Overcoming the Challenges of Migration through Employment."

[12] See also the GFMD Platform for Partnership: www.gfmd.org/pfp

- Moldova's *National Action Plan on the Protection of Children Without Parental Care (2010–2011)* and a Child Rights Information Center. The Center has produced "pocket guides" for parents and children[13] and a guide for professionals who work with children of migrants (teachers, psychologists, youth workers, NGOs). In 2005, a movie on this issue ("360 degrees of loneliness"), made by a teenage girl from Moldova, received the Award of the International Festival of One Minute Movies in Amsterdam.
- The Philippines' deployment of social workers in Japan (ISSJ) for child protection, alternative parental care, and other services for migrants and their families. A scheme also exists to provide scholarships to children of migrant workers in families remaining in the COO.
- Portugal's "Choices Program" (*Programa Escolhas*) which, since 2001, organizes projects for equal opportunities and social inclusion of the descendants of immigrants and ethnic minorities. By 2009, the Choices Program had brought together 780 institutions and 480 technicians covering about 81,695 recipients.
- Programs of the Sri Lanka Bureau of Foreign Employment were created to protect and support families of Sri Lankans working overseas. The Bureau has established a special unit to provide for the needs of the children of migrant workers, including a 24-h service to coordinate the protection and welfare of the children.[14]

Other good practices showcased by the ILO and Council of Europe include the following:

- Spain's recognition of the right to free access to education of children (up to the age of 18) of migrants in an irregular situation (Cholewinski 2006, p. 37)
- The policies of the Swiss Cantons of Geneva and Lausanne (2010) permitting children of migrants in irregular situation between the ages of 15 and 18 years to take vocational training in professional schools[15]

Participants in Roundtable 2.2 in Puerto Vallarta were concerned to see how little has been done so far to formulate specific policies to reduce risks and vulnerabilities for unaccompanied migrant children. Old vices such as under-registration of migrant children still prevail in the systems designed to address the phenomenon. Children and adolescents increase their vulnerability to being collaterally affected by migration, either by staying with members of their family left behind, or as part of a second generation of migrants in countries of destination, or by migrating with or without their parents or guardians.

---

[13] See the Child Rights Information Centre website: www.childrights.md

[14] Sri Lanka Bureau for Foreign Employment, retrieved 29 June 2010 from: http://www.slbfe.lk/article.php?article=28

[15] "Sans papiers, mais avec CFC, et le canton de Genève promet de former tous les jeunes," *Le Courrier*, March 4, 2010.

Gender Perspective, Migration, and Development

Gender has been a recurrent crosscutting theme in the GFMD since 2007. This is more than justified since despite the fact that around half the 214 million migrants in the world are women, migration policies globally have not sufficiently adjusted to this reality. In 2010, Roundtable 2.2 dealt with the gender perspective in a different way, recognizing the need to address both migrant women traveling on their own (autonomous migration) and as part of transnational families, either migrating or staying behind. This makes it possible for policies to move beyond treating "women as part of the family" only. Even when women are the principal migrants, they are also still part of transnational families, especially when they remain economically responsible for children and/or elderly left behind.

One of the papers prepared for the 2010 Civil Society Days argues that migration policies explicitly or implicitly are either (a) gender-biased, (b) gender-blind, (c) gender-neutral, or (d) gender-sensitive. They range from plain discriminatory policies to well-informed policies that stop discrimination against women at all stages of the migration process, assess which gender in the migrant family unit benefits from the policies, and what targeted measures can help them draw the best from the migration experience. Achieving formal gender equality is important (de jure), but it is only one step toward material equality (de facto). This is why independence and liberty for migrant women have been identified as two pillars of any gender-sensitive policy, together with information, education, and training in general (Schwenken 2010, p. 4).

The Committee on the Elimination of Discrimination against Women (CEDAW) considers that mixed impacts of migration on women migrant workers and families left behind are a function of (a) enabling gender-sensitive policy and institutional environments; (b) the ability of migrants and their families to access resources/entitlements; (c) gender responsiveness of cultures of countries of origin and destination, which can include family and kinship systems, community attitudes, etc.; (d) value placed on the contributions of migrants, including migrant women; and (e) levels of shared domestic responsibility between household members, especially men and women in the family (CEDAW 2008).

The main observation to be drawn from such differentiated impacts is that gender-biased, gender-blind, or gender-neutral policies may be hindering the full developmental potential of migration, both for the country of origin and destination. Continued discrimination against migrant women, or the lack of gender perspective in hiring workers from abroad while tying visas to the principal male migrant, perpetuates dependency and unfair treatment of migrant women – either accompanying the main migrant or on their own – and limits their capacity to participate in formal labor markets abroad. This can result in the curtailment of women migrants' ability to send remittances home and their capacity to take full advantage of the human development opportunities in the country of destination.

Intended or unintended gender-biased and gender-blind policies are often built on preconceived gender roles, as a function of power distribution within a society and within families, based on discrimination by sex. In particular, when combined with migration, gender bias rarely exists in isolated form, but appears interlinked with other subtypes of discrimination by race, nationality, occupation, and socio-economic status. For example, domestic violence has been recognized as an important push factor for female migration in Asia and Latin America,[16] while many migrant women start their migration experience in a disadvantaged way due to limited resources as a result of structural exclusion from formal labor markets and multiple discrimination biases in countries of origin. That is how "gender determines the different social relations which are basic to the articulation of migration and social institutions (such as family, labor markets, etc.), in the place of origin and in the place of destination of immigrants" (Hondagneu-Sotelo 2010, p. 428).[17]

Some studies have argued that in certain contexts, migration can empower women. This may even be the case where families and social networks in developing countries are strongly marked by gender bias. For example, in Oaxaca, a Southern state of Mexico, indigenous women who stay behind are able to replace their migrant husbands living in the United States as leaders of the traditional rural community, either in local posts, parents' committees, church organizations, or communitarian parties (*mayordomías*). Other observers have suggested that instead of seeing empowerment of women staying behind as a result of male migration, it may be preferable to speak about the modest gains for women due to lowered inequalities or rearrangement of gender frontiers within new scenarios (Ariza 2007, p. 488).

Three good reasons for better protecting women migrant workers were agreed during the session in Puerto Vallarta: (a) because their human rights are yet to be fully protected in practice, (b) for their contribution to development via their participation in labor markets in destination countries and by sending remittances home,[18] and (c) because they constitute a significant proportion of migrants in destination countries, slightly higher than 50% of immigrant stocks in some countries. It was acknowledged that empowering women migrants can be a critical factor for family growth, community development, and poverty reduction. But this potential can be diminished if policies in countries of origin and destination remain gender-blind. For example, many migrant women do not find proper health care in countries of destination, largely because of lack of information about health services or because of fear of deportation.

[16] See Lam and Hoang (2010), and IOM Colombia (2007).

[17] Author's translation into English.

[18] For example, there is evidence that women migrants and/or recipients of remittances are more likely to channel these earnings into the education, health, and well-being of the family than their male counterparts (World Bank 2008).

There is rapidly growing global demand for female workers, particularly care workers, due to aging populations in affluent countries.[19] Thus, the feminization of migration is most likely to continue for some time. Even if women are contracted for high-skilled and low-skilled jobs, they tend to concentrate in feminized jobs such as care, nursing, and domestic work. In particular, domestic work has attracted widespread concern because it is invisible to most government regulation and monitoring and out of sight of courts and NGOs, thus making employees especially vulnerable to exploitation, abuse, and discrimination. For example, in some countries, a woman migrant worker may become undocumented when she leaves her job, without the possibility of renewal of stay and work authorization. Such a situation can be a major deterrent for women to report abuse or mistreatment to the authorities.[20] In addition, the labor market demand for domestic workers is not recognized in most industrialized countries (e.g., Western Europe and the United States), causing large numbers of domestic workers to end up in irregular migration situations (Galloti 2009, p. 31–33).

Box 7.2 Domestic Workers Convention 2011

On June 16, 2011, the International Labour Conference of the International Labour Organization adopted the Convention concerning decent work for domestic workers, also referred to as the Domestic Workers Convention, 2011 (No. 189). The Convention offers specific protection to domestic workers. It lays down basic rights and principles and requires states to take a series of measures to make decent work a reality for domestic workers.

Domestic Workers Recommendation No. 201, also adopted by the International Labour Conference of 2011, supplements Convention No. 189. Unlike the Convention, Recommendation No. 201 is not open for ratification. The Recommendation provides practical guidance concerning possible legal and other measures to implement the rights and principles stated in the Convention.

(continued)

[19] Even if the exact number of foreign domestic workers is unknown, ILO estimates suggest that there are at least 112 million domestic workers from 29 countries worldwide. For instance, in France, over 50% of migrant women are employed in domestic work, while Italy has over 1.2 million domestic workers who are mostly migrant women. ILO LABORSTA database 2005, 2007, 2008; and ITUC (2011), available at: http://www.ituc-csi.org/IMG/pdf/ITUC_dwd_EN_BR.pdf. See also the BP and Launch Message for UNWOMEN/IOM E-Discussion "Domestic workers count too. Visibilising and Protecting Women Migrant Domestic Workers through Legal and Social Protection," retrieved March 15, 2011 from: http://www.migration4development.org/sites/m4d.emakina-u.net/files/upload/e_discussion_concept_note_final.pdf

[20] Some actions included in CEDAW (2008) to protect women migrant workers particularly in unregulated sectors are the following: (a) prohibiting the confiscation of travel documents, (b) preventing dependence on one employer, by allowing a person to change employers without risking loss of the work permit, and (c) enabling unionization of domestic workers (paragraphs 10–15).

Box 7.2 (continued)

Convention No. 189 defines domestic work as "work performed in or for a household or households." This work may include tasks such as cleaning the house; cooking; washing and ironing clothes; taking care of children, elderly, or sick members of a family; gardening, guarding the house, driving for the family, and even taking care of household pets.

Under the Convention, a domestic worker is "any person engaged in domestic work within an employment relationship." A domestic worker may work on a full-time or part-time basis, may be employed by a single household or by multiple employers, may be residing in the household of the employer (live-in worker), or may live in his or her own residence (live-out). A domestic worker may be working in a country of which she/he is not a national.

All domestic workers are covered by Convention No. 189, although countries may decide to exclude some categories, under very strict conditions.[21]

Beyond both migration and development policies, domestic work and care should be properly waged and regulated jobs urbi et orbi. In many countries, domestic work does not fall under national labor laws. This is clearly unfair even for nationals working in this sector since the lack of a written agreement can deprive them of legal protection. But it is even worse for migrant women who are usually tied to an employer, and a breach of the labor contract for any reason can result in her irregular status and be a cause for her deportation. The requirement to live in the homes of their employer can lead to abuse and exploitation of women migrant workers. Finally, as stated in the Annex of Roundtable 2.2 "Uncovering the Interfaces between Gender, Migration, Family and Development: The Global Care Economy and Chains":

> When (care work) is provided for a salary in foreign employer households by migrants, the caregiving back home in the COO is taken over by other members of the family and/or by paid lower-class domestic workers, in both cases mostly women. Women thus become integrated into gendered, occupationally segmented global care economy chains involving a hierarchy of women in CODs and COOs.
>
> (…)
>
> Although the main migration flows for domestic work have been from south (Latin America, Southeast Asia, Eastern Europe, East Africa) to north (North America, Europe and Oceania), south-south migration flows are now appearing, with demand for foreign domestic workers growing in middle income countries such as South Africa, Brazil, Chile, Argentina, Singapore and India, that have better living standards and see women's increased employment. Countries at various development levels are therefore part of global care chains and can simultaneously be COOs, CODs and countries of transit (COT).
>
> (…)

[21] Extract from ILO (2011). Governments are currently considering whether to ratify the Convention in full or with reservations/declarations or not.

The full text of ILO *Convention No. 189* can be found at http://www.ilo.org/ilolex/english/convdisp1.htm

> It is suggested that both migrant and non-migrant women in the chains may benefit from a new paradigm which challenges the role of women as unpaid caregivers. One way of achieving this is by involving men more, as they still tend to be indirect beneficiaries rather than major actors in the chains. Another good strategy is to provide schemes in CODs for proper documentation of foreign domestic workers.[22]

Migration also triggers changes in gender relations within patriarchal communities. These mostly relate to some participation of men in domestic chores and/or the division of work relating to child care. However, such changes – albeit modest – may not be permanent and may revert to the premigration situation, for example, when the family group returns to their community of origin (Boehm 2008, p.107). During the preparations for the Puerto Vallarta meeting and in the Mexico City seminar on gender and migration,[23] there was repeated discussion of the fact that children left behind usually receive more comprehensive child care if the mother stays behind, rather than the father. In the latter case, other female relatives may offer their help for child care, usually taken for granted when the woman is in charge of the family. Such an outcome may also reflect the fact that communities of origin are culturally resilient and tend to preserve traditional roles of patriarchal societies, resulting in a lack of gender equality whenever child care is involved.

A final reflection on the vulnerability of migrant women is that the mode and path of migration usually follow gender lines. Due to particular features of female migration, certain routes tend to be selected over the ones mostly used by male migrants. And the routes for trafficking in persons tend to coincide with routes for smuggling of migrant women, for example, trafficked women cross frontiers with legal and valid passports and visas, because they believe that they will work in legal and decent jobs (e.g., as waitresses, bartenders, for model agencies). They may also resort to forged documentation, which can be expensive, but their captors are willing to make such an "investment" since it will be transformed into a huge debt to be paid by means of forced prostitution of the victim.[24] Moreover, "correlations with respect to high flows for trafficking for sexual exploitation and regional reputations for corruption and organized criminal networks have been noted across the globe" (IOM 2005, p. 247).

---

[22] Some of the more outstanding good practices included in the Annex for RT 2.2 include the following: Jordan (with the support of UNIFEM) has introduced a standard, unified legally enforceable contract for domestic workers, and IOM is implementing a project on children of Ukrainian migrant women in Italy to enhance the synergic involvement of local and central authorities of both countries in counteracting the negative effects of women's migration on children left behind and on the local community. Also, the standard labor contracts for Philippine domestic workers stipulate a guaranteed basic wage, regular working hours, free transport to and from the worksite, fair grounds for termination of employment, health cover, protection of documents, and rest periods.

[23] Seminar "Promoting and Protecting the Rights of Women Migrant Workers: Partnerships for Migration and Human Development: Shared Prosperity – Shared Responsibility," coorganized by the Government of Mexico and UNIFEM (part of UN Women), September 7–8, 2010, Mexico City, Mexico.

[24] For example, in Tijuana, Mexico, it has been reported that some trafficked/smuggled women are sent to beauty parlors in order to better resemble the photographs in stolen or forged passports and visas, which is hardly a modus operandi followed by smugglers to get irregular male migrants into the United States (Vargas 2005).

Most international female workforces tend to be concentrated in sectors such as domestic service, human care, nursing, and certain forms of adult entertainment. Such sectors can tend to be unregulated, unsafe and do not offer documented opportunities for women to migrate. In some environments, the migration can occur legally, but even then women and minors are easy targets for trafficking gangs, even more so when they cross borders by nonauthorized means.

Relevant practices presented during Roundtable session 2.2 in Puerto Vallarta about migration policies with gender perspective included the following:

- The Philippines Overseas Workers Welfare Administration (OWWA) provides predeparture information to women migrants about remittance transfers, self-defense, risks of trafficking, and HIV/AIDS awareness. A Gender Focal Point is also in place in all Philippine Embassies and Consulates for counseling and legal services.
- Nepal lifted the former ban on women's out-migration and, in partnership with IFAD and UNIFEM, initiated a pilot program for managerial training and the productive investment of remittances directed at women migrant workers, targeting 1,500 beneficiaries distributed across 12 entrepreneur groups.
- Mexico has installed multisector "Gender and Migration Networks" in different states of the country, seeking to design public policies, programs, and strategies with a gender perspective.
- In Portugal, the Network of Offices for Professional Insertion of Migrants has referred thousands of migrant women to job vacancies, while a Pilot Project for the Promotion of Migrant Entrepreneurship has supported 106 business projects involving 1,462 migrant women since 2009. The Portuguese National Plan against Trafficking in Human Beings (2007–2010) has completed more than 90% of actions, while the Action Plan for the elimination of female genital mutilation (FGM) covers the training of Portuguese teachers on cultural and social aspects of FGM before being deployed in the countries of migrant origin, where they offer advice on legal support available in Portugal. In addition to this, a Support Unit for Immigrant Victims of Ethnic or Racial Discrimination provides psychological and legal assistance to immigrant women who have been victims of racial or ethnic discrimination.
- In Greece, the General Secretariat for Gender Equality in the Ministry of Interior has issued a guide on migrant women's rights in the areas of labor, health, and social security. A Greek language course tailor-made for migrant mothers promotes linguistic skills and women's capacities to help their children with school attendance. Intercultural schools and special tutorial classes are available for nonnative children with long-term residence in order to prevent dropouts or poor school achievement.
- In Spain, the Strategic Plan for Citizenship and Integration includes facilities for migrant women by means of introducing gender perspective to migration policies in all phases of the migration cycle. A law to address violence against documented and undocumented women migrant workers is currently under discussion.
- In Canada, a Gender-Based Analysis (GBA) has been employed since 2000 in order to ensure that policies relating to admission requirements, health checks,

skills recognition, and family reunification, among others, are not gender discriminatory. The Canadian International Development Agency (CIDA) also offers a guide and tools to assist policy makers abroad in addressing the connections between gender and race, ethnicity, culture, class, age, and disability.

How to Test the Main Assumptions

The main assumptions of the GFMD discussion in Roundtable 2.2 were the following.

- Special policy consideration of family unity, gender-related needs, and social welfare of migrants and families abroad, as well as families staying behind, can amplify the developmental effects of migration.
- The lack of regular avenues for migration pushes children to migrate on their own, rather than as accompanying family members, and expose themselves to criminals and predators.
- Gender-biased, gender-blind, or even gender-neutral policies perpetuate gender discrimination that curtails women migrants' ability to send remittances to the countries of origin – thus reducing their impact on family formation, community development, and poverty reduction – while limiting these women's capacity to take full advantage of human development opportunities in the countries of destination.
- Recognition of domestic and care work as a job, with full social benefits and work permit/visa schemes included, would increase human development possibilities within an economic sector occupied mainly by women and help prevent abuses of millions of migrant domestic workers worldwide.

How to test these assumptions will depend on how well the relevant variables could be operationalized and their effects measured. All tests need to be aware of the concept of development agreed for GFMD 2010, which goes beyond macroeconomic definitions to consider human development. In the first case, qualitative studies can be conducted in order to prove that migrants with direct or remote access to their usual family structure are potentially more productive, healthier, more socially adaptable, and stable than those deprived of such communication or social support. That is, family access may well be a positive externality for sectors employing overseas migrant workers. The second assumption can be tested by analyzing the background of migrant children, using qualitative or quantitative approaches. For this, there is a need for more detailed data from different countries about migrant children, in general, and about unaccompanied migrant children in particular.

The third assumption requires a battery of tests and monitoring schemes. True gender-sensitive policies are still to be constructed. Focusing on the negative impacts of gender-bias or gender-blind policies for human development of migrant women and their communities of origin seems to be a good first step.

Finally, testing the fourth assumption requires examples of relevant practices that show the economic and social potential benefits of regulating domestic and care work – subject to work permit visa schemes – worldwide. One way of doing this is by quantitative and qualitative studies on the direct and indirect contributions of domestic and care workers to the GDP as well as to family well-being, health and education, or the welfare of the elderly and levels of investment in countries of destination, particularly through the remittances sent by migrant domestic and care workers to their family and/or communities.

Recommendations

Key recommendations emerging from the Roundtable session 2.2 in Puerto Vallarta were the following:

- Promote the ratification of UN and ILO Conventions on the rights of the child and migrant workers, as well as ILO's (2011) Convention on Decent Work for Domestic Workers, and their national implementation.
- Develop gender-sensitive guidelines on policies, programs, and budgets, with appropriate indicators. The impact of migration on families should also be considered.
- Consider creating an ad hoc Working Group on Gender within the GFMD in order to move forward and monitor progress on the implementation of recommendations to develop policies with a gender perspective.

At the end of the GFMD 2010 process, some questions remained about how to further mainstream family and gender in the global discussion on migration and development and how to factor them into coherent policy making to address migration and development issues.

Migration laws and politics could be revised to ensure that they are not family or gender-blind or gender-biased. Administrative staff needs to be trained in adopting both the family and gender perspective. Specific recommendations emerging from the GFMD process for 2010 also included the following:

Family reunification opportunities that include flexible measures concerning income requirements can ensure a more productive, healthy, and adaptable foreign labor force. Thus, communication options between the worker and his/her family structure should be provided if family reunification is not available for any reason. The Tulay Project for Overseas Philippine Workers demonstrates the range of possibilities for partnerships between governments and multinational entities.

Guidelines are urgently needed for assisting and servicing children and adolescent migrants traveling on their own. The Mexican Officer for the Protection of Infancy model is one of the first steps toward this end, and it may be worth extending its use regionally and internationally. As well, other models for the protection and development of migrant children who may be future migrant workers should be created and implemented.

Since care and domestic work is vital for native families' well-being in countries of destination, governments could consider visa schemes with work permits for domestic/care workers that include legal and social security protection, based on model contracts like the ones used by the Philippines.

Government agents, the judiciary, the police, health providers, as well as workers and employers' organizations must be trained to develop effective responses to gender-based violence, discrimination, and labor exploitation directed at migrant women. For this to be a genuine gender-sensitive approach, victims of exploitation and abuse must in no way be liable to deportation, while nondiscriminatory access to health care for women migrants is essential, regardless of their migration status. The Spanish Strategic Plan for Citizenship and Integration and the Canadian Gender-Based Analysis (GBA) are two of the more advanced good practices toward true gender-sensitive migrant policies all around.

Some questions at the end of the GFMD 2010 process remain current and would require serious research follow-up, for example, how to further mainstream family and gender in the discussion on migration and development or how to factor them into coherent policy making to address migration and development issues.

Final Reflection

Not all costs in migration are economic. There are also important social costs, starting with the separation of families that characterizes the transnational family model depicted in the first part of this chapter.

Even if international migration can lead to development, especially at the local level, remittances are sent from one country to another precisely in support of families left behind. Migrants would not send money across borders to foster the local development of their home communities, unless this was to the advantage of their relatives still living there. Family ties are the main motivation for sending remittances, which can be either consumed or invested in productive projects.

Migration policies should thus take into account the social costs of migration and provide solutions to help migrants cope with the separation from their customary family structures. The GFMD 2010 meetings focused on transnational families, children traveling on their own (most looking for relatives in countries of destination), and women, both those left behind and those traveling by themselves or with their families. Each of these invokes complexities that can curtail the positive potential of migration, both for countries of origin and destination.

The main conclusion is that policies can benefit greatly from the adoption of the family and gender perspective, as illustrated by the relevant practices proposed by countries in Roundtable 2.2 of the GFMD 2010. Migrants and COOs and CODs can benefit from this perspective, basically because of the positive externalities that are fully uleashed by these means.

References

Ariza, M. (2007). Itinerario de los estudios de género y migración en México. In M. Ariza & A. Portes (Eds.), *El país transnacional: migración mexicana y cambio social a través de la frontera*. México: IIS-INM-M.A. Porrúa.

Ariza, M., & Portes, A. (Coords.). (2010). *El país transnacional: migración mexicana y cambio social a través de las fronteras*. México: IIS-INM-M.A. Porrúa.

Boehm, D. (2008). Ir y venir: historias transnacionales, trayectorias determinadas por género. In F. S. Alanís Enciso (Coord.), *¡Yo soy de San Luis Potosí!.. con un pie en Estados Unidos*. México: INM/El Colegio de San Luis/IPICYT/M.A. Porrúa.

Castro, G. L. (2010). *El Síndrome de Penélope: problema de salud en esposas de migrantes*. Paper prepared for the Puebla Process 2010 (Seminario Regional sobre migración y familia) Tijuana, Baja California, México.

CEDAW. (2008). *General recommendation no. 26 on women migrant workers*. Available at http://www2.ohchr.org/english/bodies/cedaw/docs/GR_26_on_women_migrant_workers_en.pdf

Cholewinski, R. (2006). *Irregular migrants: Access to minimum social rights*. Brussels: Council of Europe Publishing.

Fisher, J. (1989). *Mothers of the disappeared*. Boston: Southern Press.

Galloti, M. (2009). *The gender dimension of domestic work in Western Europe* (International Migration Papers No. 96). International Labour Office (International Migration Programme).

Hondagneu-Sotelo, P. (2010). La incorporación del género en la migración: no sólo para feministas ni sólo para la familia. In M. Ariza & A. Portes (Coords.), *El país transnacional: migración mexicana y cambio social a través de las fronteras*. México: IIS-INM-M.A. Porrúa.

ILO. (2011). *Factsheet, what the domestic workers convention is about*. Geneva: ILO.

IOM. (2005). *Data and research on human trafficking: A global survey*. Geneva: IOM.

IOM Colombia. (2007). *Género y remesas: migración colombiana de AMCO hacia España*. Bogotá: IOM.

ITUC. (2011). *Decent work, decent life for domestic workers: An action guide*. Available at http://www.ituc-csi.org/IMG/pdf/ITUC_dwd_AnglaisWEB.pdf

Lam, T., & Hoang, L. A. (2010, July 13). *Effects of international migration on families left behind*. Paper presented at the Experts Meeting, Civil Society Days GFMD, Mexico 2010, Mexico City.

Schwenken, H. (2010). *Gender sensitive migration policies*. Paper prepared for Round Table 2.4. of Civil Society Days, GFMD 2010.

United Nations General Assembly. (1990). *International convention on the protection of the rights of all migrant workers and members of their families*. Available at http://www2.ohchr.org/english/law/pdf/cmw.pdf

Vargas, F. (2005). *Diagnóstico preliminar sobre la trata de personas, mujeres, niñas y niños: Frontera Norte de México con particularidad en Baja California, Mimeo*. Report to IOM Mexico.

World Bank. (2008). *The international migration of women*. Washington, DC: World Bank.

Chapter 8
Textbox 4: Global Care Workers at the Interface of Migration and Development

Jean D'Cunha

Domestic work and domestic workers have increasingly come into the international spotlight. Their growing numbers, expansion of the sector, integration into global care chains, job opportunities – including the potential for upward mobility and recent advances in the normative framework for all domestic workers – the ILO Convention No. 189 on Decent Work for Domestic Workers, 2011, and others that formalize this category of informal service sector work have drawn domestic work and domestic workers from the margins to the center of the migration-development discourse and action.

The global care industry was a specific case study discussed in the Roundtable on "Gender, Family, Migration and Development" at the GFMD 2010 in Puerto Vallarta. Governments agreed to continue with this as a GFMD focus, and the Swiss GFMD in 2011 organized two workshops on "Domestic Workers at the Interface of Migration and Development" in the Caribbean and Africa, in close partnership with the governments of Jamaica and Ghana, UN Women, IOM, and other international agencies. The workshops marked the first two of a series of global outreach meetings proposed by UN Women, aimed inter alia at better understanding and implementing ILO Convention 189.

The workshops in Jamaica and Ghana reflected on the links between gender, migrant domestic work, and development and considered some good practices in (1) policy and legislation; (2) social protection, services, and self-organization; and (3) safe remittance transfers, asset building, and investments, including diaspora investment.

J. D'Cunha (✉)
UN Women, New York, NY, USA
e-mail: jean.dcunha@unwomen.org

I. Omelaniuk (ed.), *Global Perspectives on Migration and Development: GFMD Puerto Vallarta and Beyond*, Global Migration Issues 1, DOI 10.1007/978-94-007-4110-2_8,

ILO estimates that there are 100 million domestic workers globally, 83% of them women, and a large unknown number of them working overseas. The migrant domestic workers contribute their labor and expenditures to destination countries and remittances to support families and communities, at home, particularly in poorer countries. The sector is poised to grow, as global demographic and labor market developments in both high- and low-income countries generate increasing demands, mostly for women from poorer origins.

Yet 44% of countries do not regulate domestic work, which raises a number of development-related concerns for the GFMD across its full range of general thematic priorities: migrant labor and human rights, decent work, risks to personal safety and security, social and financial protection, skills and job matching, transnational families, gender equality, best interests of the child, costs of migration, migrant health, and the acute vulnerability of undocumented migrant workers.

The GFMD 2011 workshops recognized that governments need to create an enabling socioeconomic, legal environment to ensure safe and regular employment of domestic workers, for example, through tax credits for households and schemes that increase the affordability and quality of care in the home, and create more and decent work for domestic workers. Participants suggested ratifying and implementing the ILO Convention 189 and accompanying instruments[1] by working in multistakeholder partnerships to amend or create policies and programs to protect domestic workers. These would be supported by stronger evidence and data, capacity building of officials, public consciousness-raising, enforcement of regulations, and effective monitoring and evaluation.

Some promising practices already exist, including Jamaica's National Minimum Wage Act, legislation on employment and termination, and legislation on trafficking in persons that protect domestic workers. In the USA, the New York Domestic Workers Bill of Rights protects domestic workers, including undocumented workers. In Ghana, a multistakeholder task force has been set up to formulate policy, review existing labor legislation protecting domestic workers, and develop a national framework to ensure decent work for domestic workers and support ratification of ILO Convention 189. South Africa was reviewing its minimum wage for domestic workers and discussing a bilateral agreement with Lesotho to regularize undocumented migrant workers, including migrant domestic workers.

Regional processes can also (but seldom do) facilitate free and safe cross-border movement of domestic workers. The Caribbean Single Market

[1] ILO Convention No. 189 on Decent Work for Domestic Workers (adopted on June 16, 2011), the CEDAW General Recommendation No. 26 on Women Migrant Workers and the Migrant Committee's General Comment on Migrant Domestic Workers. These instruments are treated as complementary and mutually reinforcing, given their similarities, albeit different emphases, processes, constituencies, and levels of ratification.

Economy (CSME) is promising in that it includes domestic workers as a skill category in the free movement of labor since 2009, but this has not worked in practice, because of a lack of awareness among migrants about their rights and entitlements under the CSME and the different levels of certification in the region. At the Ghana workshop, participants urged that gender concerns, particularly of migrant domestic workers, be mainstreamed into the African Decade for Women Action Plan, the ECOWAS Gender Action Plan, the East African Community, and the SADC agreement on free movement of people.

Governments and their non-state partners agreed in Jamaica and Ghana that access to frontline social and financial services would empower domestic workers economically and socially. Health insurance smart card schemes were an option, as well as portability of social security, family support services to offset the social costs of migration, unemployment insurance and maternity benefits, information and orientation, skills training, services to address abuse, and mobilization and training for collective organizations.

Since women remit a higher percentage of income than male migrant workers and tend to invest it in the health, education, and well-being of their families, governments and private sector partnerships could leverage women's remittances and savings and facilitate asset building and investments in ways appropriate to women domestic workers' needs. In addition to the general need to reduce remittance transfer costs and promote financial literacy as Women's World Banking was already doing, microfinance institutions needed to be empowered to handle foreign currency, loans, and credit and help increase female domestic workers' access to credit, loans, and investment options.

One of the most concrete outcomes of the two workshops was an agreed Checklist to Protect and Support Migrant Domestic Workers. This is intended to be used as a practical guide to governments embarking on new or revised policies, legislation, and programs to implement ILO Convention 189 and accompanying instruments. In Jamaica, it was also proposed to establish regional networks of domestic workers and representative groups to help move forward the ratification and implementation of ILO Convention 189 and accompanying instruments, launch sensitization programs on these instruments, and ensure domestic workers are part of the policy decision-making and actions.

Regarding financial empowerment, the Ghana workshop recommended pilot partnerships among diaspora, financial institutions, migrant domestic workers, policy makers, and employers to develop schemes for workers to pool their resources and access financial services and products tailored to their needs and financial interests. To strengthen the knowledge base on this phenomenon, it was suggested that regional Migration Profiles be prepared (or existing Migration Profiles revised) on cross-border domestic work. An international (or region-wide) body could also be set up to oversee the regulation and licensing of migrant recruitment agencies, and that recruitment agencies are appropriately trained on the rights and the roles of domestic workers.

Chapter 9
Irregular Migration: Causes, Patterns, and Strategies

Stephen Castles, Magdalena Arias Cubas, Chulhyo Kim, and Derya Ozkul

Introduction

The vast majority of the world's 214 million international migrants have regular immigration status. Estimates suggest that only between 10% and 15% of migrant populations are in "irregular" status (IOM 2010b, p. 29). *Irregular migration* takes place when a person enters or lives in a country of which he or she is not a citizen, in violation of its immigration laws and regulations. Many irregular migrants are workers, but not all enter the labor market: people migrating to join family members or for lifestyle reasons may become irregular. Rejected asylum seekers may also fall into irregularity. While irregular migration affects most regions of the world, statistics tend to be unreliable, due to the very nature of such movements. Estimates are often manipulated for political reasons, since blaming irregular migrants for social problems has become a widespread practice of populist politicians. In addition, numbers of irregular migrants can vary for economic reasons (e.g., the global economic crisis of 2007–2011 affected patterns) but also due to changes in laws.

Irregular migration is complex and hard to define (Düvell 2006). Many migrants enter a country legally but become irregular by staying longer than permitted or by working without permission. Some migrants pass in and out of irregular status (Reyneri 2001). Others are "semi-compliant": they seek to conform to national laws and rules, but are unable to do so fully (Martin Ruhs and Anderson 2006). As the International Organization for Migration (IOM) points out: "The category of semi-compliance – the 'space between' strict legality and (il)legality – is extremely broad…" (IOM 2008, p. 203). Many irregular migrants strive to become regu-

S. Castles (✉) • M.A. Cubas • C. Kim • D. Ozkul
Department of Sociology and Social Policy, University of Sydney,
Sydney, NSW, Australia
e-mail: stephen.castles@sydney.edu.au

I. Omelaniuk (ed.), *Global Perspectives on Migration and Development: GFMD Puerto Vallarta and Beyond*, Global Migration Issues 1, DOI 10.1007/978-94-007-4110-2_9,

lar, and some governments facilitate this through "regularization," "legalization," "amnesty", or "registration" programs (Castles and Miller 2009, pp. 181–186). In other cases, migrants try to get closer to regular status by buying or even hiring false documents (Vasta 2011). Rather than a dichotomy of regular and irregular status, it would be more accurate to see migrants as being somewhere on a *continuum* from regularity to irregularity.

Other terms are often used in political and media discourses, such as "illegal," "undocumented", "unauthorized", or "clandestine" migration (Clandestino 2009a). However, public perceptions of irregular migration have become highly politicized, so that terms may be value-laden and often negative. In this chapter, the term *irregular migration* will be used as the most neutral descriptor. "Undocumented migration" is also a relatively neutral term. In principle, it is wrong to speak of "irregular migrants" – no person is "irregular" since all have human rights and dignity that should be respected.

Clearly, irregular migration is a complex phenomenon. This chapter provides an overview of causes, types, and available data for various world regions, including case studies of Mexico, Italy, Turkey, the Republic of Korea, and Malaysia. It discusses initiatives to achieve improved policy responses and concludes by discussing the relationship between irregular migration and human development.

What Causes Irregular Migration and Residence?

The causes of irregular migration and residence are manifold and cannot be dealt with exhaustively here. We discuss some of the main factors and draw attention to some others.

National Laws and Regulations

If we lived in a world of freedom of movements – a demand both of liberal economists (Bhagwati 2003; Straubhaar 2002) and human rights advocates (Harris 2002; Pécoud and de Guchteneire 2007) – there could by definition be no irregular migration. *Open borders* would be the simplest way of eliminating irregular migration – but political considerations hinder governments from adopting this remedy. Irregularity can therefore be seen as a consequence of laws and regulations, which label certain forms of mobility as legal and desirable, and others as illegal and unwanted.

A state's right to control entry to its territory and to decide who may be a citizen is often seen as an enduring aspect of national sovereignty. However, concern with "illegal migration" is a quite recent phenomenon, connected with the tightening up of border controls and passport regulations in the twentieth century. According to Düvell (2006, pp. 21–29), the concept of illegal migration emerged mainly after the abandonment of migrant labor recruitment schemes by European nations in the mid-1970s.

The politicization of migration and asylum became dramatic after 2001: some section of the media and right-wing politicians portrayed "uncontrolled migration" as a threat to security. As a result of the "securitization" of migration (Bigo and Guild 2005; Guild 2009), the public often has highly inflated ideas of the volume of irregular migration, while centrist and left-wing politicians fear that they will be seen as "soft" on migration if they do not take strong control measures.

Modern states do not treat all migrants the same, but instead select and differentiate. *Selectivity* means privileging certain people as suitable for entry and residence and rejecting others. For instance, until the 1960s, the USA, Australia, and Canada excluded non-Europeans. Today, regional organizations – such as ECOWAS in Africa and MERCOSUR in Latin America – have preferential rules for admission of nationals of member states. The European Union (EU) goes furthest with freedom of movement of EU citizens to other member states, combined with strict regulation of entry of "third country" citizens. *Differentiation* refers to discriminating between potential entrants on the basis of economic criteria such as education and skills ("human capital"), type of migration, or gender. Virtually all industrialized countries have preferential rules to attract highly skilled migrants for the information technology (IT), engineering, scientific, and medical sectors. At the same time, such states claim that they have little need for lower-skilled workers. Yet employer demand for such workers remains strong (CEC 2005).

A central issue at the national level is thus the *contradiction between state and market*: often migration rules signal migrants to stay out, while the market signals that they are welcome. Governments sometimes use "crackdowns" on irregular migrants as a way of appeasing public opinion, while tacitly permitting irregular labor migration to meet employer demand. Some employers actually prefer irregular migrants, because they lack rights, cannot complain to authorities or trade unions, and are therefore easily exploitable.

Similarly, many states have restricted entry for asylum seekers, despite being signatories to the United Nations (UN) 1951 Refugee Convention and its 1967 Protocol. This creates another paradox: asylum seekers have a right to apply for asylum, but cannot enter a country legally to do so, and therefore have to enter in an irregular way, sometimes using people smugglers. The problem is particularly acute in the case of "mixed flows," such as irregular movements of Africans across the Mediterranean to EU countries. The United Nations High Commissioner for Refugees (UNHCR) has issued special guidelines for differentiating between economic and forced migrants, while respecting the human rights of both (UNHCR 2006).

Differentiation on the basis of gender is also common. Britain in the 1970s subjected Asian women entering for marriage purposes to virginity tests (Lutz et al. 1995). Open gender discrimination is less frequent today, although Singapore still makes migrant women workers take pregnancy tests and deports those found to be with child, while other Asian countries admit migrant women specifically for marriage to local men (Bélanger et al. 2010). In other cases, women are preferred for domestic service (Lutz 2008), lower-level health service jobs, entertainment, and other forms of poorly paid and exploitative work. Irregular migration is very common in occupations seen as "typically female."

Neoliberal Globalization

Globalization creates the conditions for increased human migration in two ways:

> *First*, it provides the technological and cultural basis for mobility. Electronic communications diffuse knowledge of migration routes and work opportunities. Long-distance travel has become cheaper and more accessible. Once migratory flows are established, they generate "migrant networks": previous migrants help members of their families or communities with information on work, accommodation, and official rules. "Transnational communities" emerge: people have a sense of belonging to two or more nation-states and have economic, social, political, and cultural relationships that go across border (Portes 1999). This may encourage irregular migration: while governments remain focused on national control models, migrants may follow the logic of globalized labor markets and transnational relationships.
>
> *Second*, the neoliberal mode of globalization since the late 1970s has led to vast increases in inequality both between the Global North and South, and within countries and regions. Global inequality by the mid-2000s was "probably the highest ever recorded" (Milanovic 2007, p. 39). Another form of inequality is the gulf in human security between North and South, which finds its expression in poverty, hunger, violence, and lack of human rights in the South. As the Global Commission on International Migration (GCIM) pointed out: international migration is driven by "development, demography and democracy" (GCIM 2005, p. 12).

Irregular migrants help provide the labor market *flexibility* central to neoliberal globalization: they can be easily hired and fired. Manufacturing production processes can often be divided, so that high-productivity and high-skilled activities are carried out in advanced economies, while low-productivity, low-waged segments are outsourced to the South (Delgado Wise 2007). However, where economic activities must be carried out close to their consumers – especially construction and service tasks – labor market deregulation has opened the way for trends toward informalization. In Africa, Latin America, the USA, and parts of Europe and Asia, the informal (or "shadow") economy is a major sector, where migrant workers often suffer exploitative work relationships (Jones 2000; Martin 2004; Mingione and Qassoli 2000; Reyneri 2001; Sadiq 2005). Even in countries with strict immigration control – such as the UK, Germany, and Japan – the growth of subcontracting, casual employment, and temporary employment reflects an expansion of the informal sector (Martin et al. 2006; Ruhs and Anderson 2006).

The Individual and Collective Agency of Migrants

Official policies on irregular migration are generally based on the idea of migrants as economic beings. They ignore the social relationships of migrants as members of families and communities, as well as the way personal characteristics and goals change over the life cycle. Anthropologists and sociologists use the concept of *migrant agency* to analyze the ways in which migrants actively shape migratory processes.

Migration decisions are often made not by individuals but by families. In situations of rapid change, a family may decide to send one or more members to work elsewhere, in order to maximize income and survival chances. Remittances sent home by irregular migrants can help lift families out of poverty and may contribute to investments and economic development. Family linkages often provide both the financial and the cultural capital (i.e., the knowledge of opportunities and means of mobility), which make migration possible. Motivations change over migrants' life cycle: in economic migration, the primary migrant is usually a young man or woman in search of temporary work, who often intends to return home once certain savings targets have been reached. Difficulty in achieving such targets may lead to prolonged stay. This in turn encourages family reunion or formation. People start to see their life perspectives in the new country. Once migrants' children go to school, learn the language, join peer groups, and develop bicultural or transcultural identities, it becomes very difficult for the parents to leave.

Migration laws and rules often ignore the social nature of the migratory process, and fragment communities into bureaucratic categories. These categories may not correspond to social realities. In such cases, irregular migration and residence are likely to arise.

The Activities of the "Migration Industry"

As so often in contemporary life, if the rules imposed by governments and bureaucracies become too complex, it may be necessary to seek professional help. The migration industry includes migration agents, travel bureaus, bankers, lawyers, labor recruiters, interpreters, and housing brokers. It may also include shopkeepers, priests, and teachers, who help their compatriots on a voluntary or part-time basis. Facilitating migration is a major and mainly legal international business (Salt and Clarke 2000, p. 327). But there is sometimes no clear division between organizations providing legitimate recruitment and travel services, and those indulging in people smuggling or trafficking. The migration industry may become one of the driving forces behind irregular migration when government policies are altered. The forms of migration may change – for instance, from legal worker recruitment to asylum migration or undocumented entry – but the volume may be undiminished.

Other Causes

People may become irregular migrants for a variety of other reasons. Sometimes state borders move or new states are established, and parts of the population may find their legal status changed. This was the case after the collapse of the Soviet

Union, when former Soviet citizens suddenly became non-nationals of newly created states. Others become irregular when they flee conflict or persecution, but fail to secure recognition as refugees. Yet others can become irregular when state rules on residence or citizenship change, as happened to the *sans papiers* in France in the 1990s. These and other causes cannot be discussed in detail here.

Types of Irregular Migration

A widespread popular image of irregular migrants is of people crossing the sea in small boats, or walking through deserts and climbing fences. Such illegal entry does take place but affects only a minority of irregular migrants; most enter legally, but then take up residence or employment without the necessary permits. The public is fixated on "boat people", who seem to evoke archaic fears of invasion, while remaining oblivious to the much larger number of legal entrants arriving by air, who drift into irregularity. In Australia, for example, there was media outrage over 4,000–5,000 asylum seekers arriving by boat in the first half of 2010, while almost no attention was paid to the much larger number of irregular residents – an estimated 48,900 in mid-2009 (DIAC 2009).

The International Centre for Migration Policy Development's (ICMPD) Clandestino project (Clandestino 2009a) differentiates between irregular foreign residents and irregular foreign workers. The former live in EU countries without official permission, while the latter work without official permission. There is an overlap between the two categories, as Clandestino's list of the main types of irregularity shows:

Irregular foreign residents who are not irregular workers

- Seemingly regular registered foreign nationals with falsified papers
- Children without residence status
- Aged family members without residence status

Irregular foreign residents who are also irregular foreign workers

- Foreign nationals without residence status in regular tax-paying jobs
- "Tourists" from non-EU countries in irregular unregistered jobs
- Foreign nationals without residence status in irregular unregistered jobs

Irregular foreign workers who are not irregular residents

- Foreign nationals with residence status but without work permission in irregular unregistered jobs
- Foreign nationals with residence status and work permission in irregular unregistered jobs

If the reference to "EU countries" is removed, these categories may be applied globally. However, the Clandestino project does not cover people smuggling or trafficking. It is therefore necessary to add the following categories (based on ILO 2006; IOM 2010a):

- Smuggled migrants: persons who use people-smuggling organizations to obtain illegal entry to a state. People smugglers may provide transportation or guidance to cross a single border (e.g., from Morocco to Spain, or Mexico to the USA) or organize long-distance migration across many borders (e.g., from Pakistan to the UK via Iran, Turkey, Greece, Italy, and France). People-smuggling organizations sometime include former migrants and corrupt officials, as well as middlemen along the route. Smugglers may be motivated by the desire to help migrants, as well as by profit.
- Trafficked migrants: persons who are transported across international borders, often against their will, for the purpose of exploitation and forced labor. Trafficking may involve forcing both men and women but applies particularly to providing women and children for the sex industry. Trafficking often involves organized criminal gangs. Trafficked migrants may be detained against their will in the destination country, forced into debt bondage or prostitution (or both), and subjected to sexual assault and other forms of violence.

Legal definitions of people smuggling and trafficking are provided in the Vienna Protocols to the United Nations Convention against Transnational Organized Crime of 2000 (UNODC 2004). The Palermo Protocol to Prevent, Suppress and Punish Trafficking in Persons, especially Women and Children (2003) has been signed by 117 countries (UNODC 2011). IOM stresses the importance of distinguishing between people smuggling and trafficking (IOM 2010a). Smuggled migrants are buying a service from the smugglers, and their relationship should end once they arrive in the destination country. Trafficked persons, by contrast, face long-term bondage and exploitation. However, both smuggled and trafficked persons run the risks of imprisonment, deportation, violence, and even death while trying to cross borders illegally. Moreover, smuggled migrants and their families may end up with large debts to criminal organizations. The leaders of smuggling and trafficking organizations are rarely apprehended, while more subaltern members of the gangs may end in jail.

Regional Overviews on Irregular Migration

Statistics about irregular migration "vary widely and are usually imprecise" (IOM 2010b, p. 29). Since irregular migrants are usually not registered by the authorities, figures are at best estimates based on coherent methodologies and are at worst mere guesses. This absence of hard data opens the door for the wildly inflated accounts disseminated by some sections of the media. In any case, the distinction between regular and irregular migration is far from clear in many regions of the world. The best way to gain a rough picture is to look at regional and national data. Some brief case studies in boxes below illustrate the situation better.

North America

A study by the Pew Hispanic Center estimates the March 2010 unauthorized population at 11.2 million (Passel and Cohn 2011) – almost 4% of the total US population. According to an earlier Pew report (Passel and Cohn 2009), in 2008 about seven million Mexicans made up 59% of irregular immigrants. Significant regional sources of unauthorized immigrants included Asia (11%), Central America (11%), South America (7%), the Caribbean (4%), and the Middle East (less than 2%). The largest concentrations of irregular residents were in California, Texas, Florida, and New York. Most lived with spouses and children, often with US-born children who held US citizenship. An estimated 7.8 million irregular migrant workers made up 5.1% of the total US workforce. Irregulars were 10% or more of the workforce in California, Arizona, and Nevada. They were overrepresented in low-skilled jobs. Some 25% of farm workers and 17% of construction workers in the USA were undocumented.

The unauthorized population had fallen to 11.2 million in 2010 from a peak of about 12 million in 2007 (Passel and Cohn 2011). The sharp fall in Mexican migration to the USA from 2006 onward has been attributed to the economic crisis, which hit the construction industry strongly (Ruiz and Vargas-Silva 2009). Irregular migration is particularly sensitive to economic conditions, since irregular workers are not entitled to unemployment pay or other social benefits (McCabe and Meissner 2010). However, other possible factors include increased enforcement, high risks due to criminal activities, and changing demographic and economic conditions in origin countries.

It is hard to find accurate data on irregular migrants in Canada. One estimate puts their number at about 200,000 in 2006, mainly employed in construction and other blue-collar jobs (IOM 2008, p. 213).

Box 9.1 Mexico

Mexico is well known as a country of emigrants. Currently, about one-tenth of the total Mexican population lives in the USA, more than half with irregular status. Migration is a key issue of politics and international relations for both countries (Délano 2009). An immigration reform that would have legalized millions of irregular immigrants in the USA collapsed after legislation failed in Congress in 2006 and 2007 (Alba 2010). Critics highlight that the failure of US lawmakers to reform immigration law violates basic human rights principles (Jimenez 2009; Human Rights Watch 2010b): "the current system has created a massive underground of persons who have lived undocumented in the U.S. for many years" (Human Rights Watch 2010b, p. 618). The US Office of Immigration Statistics estimates that out of the 2010 total population of irregular migrants, 43% had arrived in the USA in the 1990s, and an additional 19% had entered the country during the 1980s (US Department of Homeland Security 2011). This means that more than 6.6 million people have been living as irregular migrants inside the USA for over a decade.

(continued)

Box 9.1 (continued)

In addition, increased US border enforcement made it more difficult and costly for new migrants to cross the border. One of the consequences of such approach has been an increment in the number of migrant deaths as migrants take greater risks to reach the USA (Cornelius 2001; Jimenez 2009). A recent report puts the number of migrants who died in unauthorized border crossings from 1994 to July 2009 at between 3,861 and 5,607 (Jimenez 2009). Moreover, the development of anti-immigration laws in Arizona, Utah, and other US states has seen irregular migrants increasingly racially profiled, criminalized, and deprived of basic rights (Gargis 2011; Migration Dialogue 2011).

Mexico is also a transit and destination area. Central Americans have transited through the country to seek asylum or look for jobs in the USA for decades. The Mexican government estimates that 150,000 irregular migrants enter the country each year, but NGOs believe the actual number to be much higher (Comisión Nacional de los Derechos Humanos 2011). The Mexican government has been under pressure to control its southern border to reduce irregular migration to the USA via its territory (Alba 2010). In 2010, the Mexican authorities deported more than 65,000 irregular migrants, most of whom were from Central America (Instituto Nacional de Migración 2010). The massacre of 72 Central and South American migrants by an armed gang in August 2010 in the state of Tamaulipas brought the issue of irregular migration to public attention. This case reflects an increasingly dangerous situation for migrants trying to reach the USA via its southern neighbor. Mexico's prioritization of the fight against drug trafficking and organized crime has led to a massive increase of human rights abuses, particularly in the border states. Organized crime has taken over territories through which migrants frequently travel, thus exposing them to an ever-greater risk in the face of kidnapping and extortion (Vera López 2010; Human Rights Watch 2011). Mexico's National Commission of Human Rights estimates that more than 9,000 irregular migrants were kidnapped in Mexico between September 2008 and February 2009 (Comisión Nacional de los Derechos Humanos 2011). Yet, migrants are also often victims of abuse through corruption among Mexican officials (Amnistía Internacional 2010).

The Mexican government has been severely criticized for its lack of protection of irregular migrants. In 2008, the government reformed its laws in order to decriminalize irregular migration and the provision of assistance to irregular migrants (Amnistía Internacional 2010). Mexico promulgated a new Migration Law in May 2011 after considerable public debate. This law recognizes the rights of migrants – irrespective of their legal status – to basic services such as health, education, and access to justice (Honorable Congreso de la Unión 2011). However, critics point out that the law by itself – without significant institutional reform – will not ensure an end to the systemic violation of migrant rights (Sin Fronteras 2011).

Central America and the Caribbean

Data on irregular migration for this region is far less comprehensive and reliable than for North America. Many people seek to migrate to the USA using Mexico as a transit country (see Box 9.1). About half the 500,000 migrant workers in Costa Rica have irregular status, many coming from neighboring countries like Nicaragua. As elsewhere, they are concentrated in agriculture and other low-skilled jobs, such as construction, tourism, and domestic work. In the past, many Haitians have sought to travel to the USA, often using small boats to reach Florida. Enhanced US border protection measures have led to a decline in such movements. Traditionally, large numbers of Haitians have gone to the Dominican Republic to work in the sugar plantations. Today, between 500,000 and 700,000 Haitians are thought to live in the Dominican Republic, mostly with irregular status. Another destination is the Bahamas, where 40,000–50,000 Haitians are believed to live (IOM 2008).

South America

Historically, South America was a region of immigration from Europe. With uneven economic growth from 1945, intracontinental movements increased. Then, following the economic slowdown of the 1980s, out-migration from the continent to North America and Europe grew. There was little regulation until recently, and most flows were spontaneous. For example, labor flows from Bolivia, Paraguay, and Chile to Argentina emerged from the 1940s to the 1960s. Families soon joined single and mainly male migrants, creating neighborhoods of irregular immigrants in some cities. Today, Argentina remains a country of both immigration and emigration. Most irregular migrants work in domestic service, construction, and in textile factories. Venezuela has also experienced a great deal of irregular immigration. By 1995, it was estimated that there were two million irregular residents, most of them Colombian. In addition to the oil industry, agriculture, construction, and other industries attracted migrants. However, as a result of political and economic instability in recent years, emigration from Venezuela became significant, with increased flows to the USA and Spain (Castles and Miller 2009, pp. 170–178).

Out-migration from South America to North America and Spain is also mainly irregular. The largest flows have been from Ecuador, Peru, Brazil, and Argentina (IOM 2008, p. 213). However, Spain has set up regularization procedures, often giving preference to people of Spanish origin over African immigrants. The global economic crisis has led to declines in mobility to the USA and Europe, but no slowdown in flows of migration within the continent (Mazza and Sohnen 2010).

Europe

The Clandestino database has shown that irregular migration is not as extensive as often claimed: the 2008 estimates for all irregular residents in the EU15 (the EU countries prior to the 2004 and 2007 extensions) ranged from 1.8 to 3.3 million. Irregular residents made up less than 1% of the EU's total population (Clandestino 2009a, Table 1). Estimates for some of the largest EU countries in 2008 were as follows: the UK, 417,000–863,000 persons; France, 178,000–400,000; Germany, 196,000–457,000; Poland, 50,000–300,000; Italy, 279,000–461,000; and Spain, 280,000–354,000 (Clandestino 2009a, p. 5). The wide ranges in the estimates were a reflection of the lack of exact knowledge. The number of irregular residents declined substantially between 2002 (when the range of estimates was 3.1–5.3 million persons) and 2008. The main reasons for the decline were the 2004 and 2007 EU expansions that turned many irregulars (such as Polish workers in the UK and Ireland) into legal residents and regularization programs in certain countries – notably Italy and Spain – which helped some 1.8 million persons to move into regular status in the EU as a whole.

The vagueness of the data permits only limited conclusions on the characteristics of irregular residents. In some countries, irregular residence is dominated by one country of origin (e.g., Albanians in Greece or Ukrainians in Poland). In other countries (e.g., the UK and Germany), irregulars are of many origins. Men are overrepresented among irregular residents, but the numbers of women and children are increasing. Clandestino gives no information on irregular employment, but other sources indicate the predominance of low-skilled employment, high levels of irregular and precarious work, and concentration in agriculture, construction, and lower-skilled service occupations (Berggren et al. 2007; Jordan and Düvell 2002; Reyneri 2001; Ruhs and Anderson 2006). Recruitment of irregular migrant women for domestic service has led to a revival of this sector at a time when it seemed to be disappearing (Anderson 2000; Lutz 2008).

Box 9.2 Italy

Irregular migrants enter Italy by various routes: the Italian-Slovenian border is used by those from Eastern Europe, Central Asia, the Middle East, the Indian subcontinent, and East Asia; the Italian-French border is used by migrants from Africa; while others coming through North Africa arrive by boat on the southern coasts (Clandestino 2009b). Italy has attracted irregular migrants due to its large informal economy, as well as government policies, which accepted irregular migration combined with regularization schemes as a way of meeting labor market demand. Between 1986 and 2009, six regularization schemes were introduced, which received approximately 1.7 million applications (Network Migration in Europe 2010).

(continued)

Box 9.2 (continued)

When the center-right coalition came to power in 2001, immigration was redefined as an issue of security. The Immigration Law of 2002 (Bossi-Fini Act) made it compulsory to sign a contract with an employer before migrating. Yet in cases such as domestic work and care services, this has proved to be unrealistic, as employers want to meet their employees before they offer them employment. In the same period, the government also introduced a large regularization program: around 700,000 migrants were regularized in 2002 (Pastore 2008). The Italian Navy was given the right to check ships at sea for irregular migrants. The practice of "push-backs" (*respingimento*) by the Navy has forced boats back to Libya, which is not a signatory to the 1951 Geneva Refugee Convention. Critics have argued that this policy aggravated "a growing humanitarian crisis of deaths at sea due to sinking migrant smuggling vessels" (Global Detention Project 2009).

The third Berlusconi center-right coalition introduced the "Security Package" (*Pachetto Sicurezzo*) in 2008. This included fines between 5,000 and 10,000 euros and immediate expulsion, confiscation of property and severe fines for landlords renting to irregular migrants, and requirement of a residence permit for any civil act including marriage and registering a child (PICUM 2010–2009). The initial package also suggested that health and education workers should report irregular patients and students to authorities, and that the possible detention period should be extended from 6 months to 3 years (PICUM 2010).

A key issue in Italy is the effect of the "Arab Spring". The flight of people from oppressive regimes has been portrayed as an invasion of Europe. As soon as the first boats arrived at the island of Lampedusa, Italy declared a state of emergency on the island and asked for help from other EU Member States. By mid-April 2011, over 23,000 people, mainly from Tunisia, had arrived by boat in Lampedusa. These numbers are small compared to the 460,000 people who have fled from Libya to neighboring countries. On April 6, 2011, Italy reached an agreement with Tunisia stating that a 6-month temporary residency permit will be granted to Tunisians already in Italy, while Tunisia agreed to simplify and accelerate return procedures for newly arriving Tunisians. Following the agreement, France was concerned that Tunisians who were granted a 6-month temporary residency permit could travel freely in the Schengen zone, including to France. Although criticized by the European Commission, France reinforced its border controls with Italy (Council of Europe 2011).

Middle East

It is impossible to provide an adequate discussion here of the complex migratory trends affecting the Middle East (for an overview, see IOM 2010b, pp. 205–218). The rapid growth in prosperity of the oil-producing regions of the Arabian Gulf and Libya after 2003 led to major intraregional flows of migrant workers (e.g., from Jordan) as well as interregional flows from Asia and sub-Saharan Africa, mostly in the form of contract labor. However, irregular migration is also significant, while contract workers may fall into an irregular situation if they are unable to meet strict control conditions, such as those imposed in many Gulf countries through the *Kafala* (sponsorship system). Reports of precarious conditions, absence of rights, and mass expulsions have proliferated in recent years, while the feminization of migration resulting from the growth of domestic service is also contributing to fragile and vulnerable conditions for many labor migrants (Castles and Miller 2009, pp. 164–167).

Box 9.3 Turkey

It has been estimated that there are between 500,000 and one million irregular immigrants in Turkey (Düvell and Vollmer 2009, p. 11). Turkey experiences diverse flows of irregular migration. Some are related to Turkey's geographic situation between Asia and the EU, others to the wars and political upheavals taking place in the region.

Since the collapse of the Soviet Union, significant numbers of migrants – primarily women – have arrived from Moldova, Romania, Ukraine, and the Russian Federation. Most come as tourists and become irregular by overstaying. They are mostly involved in the "suitcase trade," or work as translators, domestic workers, or in the entertainment sector (Eder 2007). They are often subject to fraud and sexual abuse (Erder and Kaska 2003; Bloch 2003). The word *Natasha* has become a popular term in Turkey used as a synonym for women working irregularly in the sex industry.

Between 1995 and 2009, nearly 58% of apprehended irregular migrants were assumed to be in transit to the EU: mainly from Iraq (129,000), Pakistan (70,000), Afghanistan (59,000), Iran (28,000), Palestine (25,000), Bangladesh (22,000), Somalia (22,000), Mauritania (15,000), Syria and Burma (over 10,000) (Icduygu 2011, p. 5). Trafficking and smuggling play a big role. Some transit migrants are able to reach the EU after a perilous journey across the Aegean; others stay in Turkey while waiting to save enough money. Irregular transit migration became very important in relation to Turkey's admission application to the EU.

(continued)

Box 9.3 (continued)

There is no comprehensive body of legislation for managing irregular migration in Turkey. However, significant new measures have been introduced as part of the "Europeanization process." Turkey made substantive changes to the Penal Code within the context of an EU reform package (accepted in 2002) criminalizing human smuggling and trafficking. Other measures were introduced to provide support to victims of trafficking (Ozcurumez and Senses 2011, p. 245).

The Turkish government believes that irregular transit migration should be managed in the context of international burden sharing (Republic of Turkey Ministry of Foreign Affairs 2011). Turkey has signed readmission agreements with Greece, Syria, Kyrgyzstan, Romania, Ukraine, Pakistan, and Russia and is in negotiations with several other countries. Since the early 2000s, a possible readmission agreement with the EU has been discussed. Turkish authorities, however, say that they will not sign the final agreement unless the EU offers visa facilitation for Turkish citizens.

Turkey is a signatory to the 1951 Geneva Refugee Convention, yet it still retains a "geographical limitation," that is, asylum seekers from non-European countries are to be resettled in other countries. This system has long waiting periods, particularly since 2003 when the numbers of asylum seekers started to increase rapidly due to human rights violations in Iraq, Palestine, and Somalia. While waiting for their situation to be clarified, most asylum seekers become irregular migrants in Turkey. Lately, the issue has gained increasing importance due to movements from Tunisia, Egypt, and Syria.

Civil society has worked to meet the urgent needs of irregular migrants for shelter, food, and legal consultation, particularly in big cities. Examples include the works of the Anatolian Development Foundation (ADF), the International Catholic Migration Commission (ICMC), the Helsinki Citizens' Assembly, Caritas and the Humanitarian Relief Foundation (IHH). Migrant networks such as Iraqi Turks Culture and Assistance Association and Afghan Turkmens Social Assistance and Solidarity Association include initiatives coming from migrants themselves. Civil society's involvement has been highly important in Turkey but has been hampered by lack of material resources in face of the massive problems.

Africa

Knowledge on irregular migration in Africa is particularly weak. This reflects both poor data collection and the predominantly spontaneous nature of migration. Important migration systems have evolved, centering on areas of economic growth such as Libya in the North, Côte d'Ivoire, Ghana and Gabon in the West, and

South Africa and Botswana in the South (Bakewell and de Haas 2007, p. 96). Trafficking of women and children for sexual exploitation and bonded labor is rife (Abella 2005).

West Africa is often seen as the most mobile part of the continent. A UN study showed an international migrant population of 6.8 million (2.7% of W. Africa's total population) in 2000 (Zlotnik 2004). The largest movements are from the northern inland to the southern coastal regions, and especially to fast-growing cities like Lagos, Dakar, and Accra. In the 1950s and 1960s, many migrants from Togo and Nigeria were attracted to Ghana. After the 1966 coup in Ghana, the government ordered a mass expulsion of some 200,000 migrants, mainly Nigerians. Due to Nigeria's oil wealth after 1973, millions of Ghanaians and other West Africans sought work there. But an economic crisis followed, and in 1983–1985, an estimated two million West Africans were deported, including over one million Ghanaians (Bakewell and de Haas 2007, p. 104). From the 1980s, West African migration patterns changed dramatically due to economic decline and civil wars (Bakewell and de Haas 2007). Mass flows of refugees and internally displaced persons took place, and labor migration patterns were disrupted. Today, multidirectional patterns of labor migration within the region persist. However, increasing numbers of West Africans, both highly skilled and less skilled, now seek work outside the region.

Post-apartheid South Africa draws in migrants from the rest of the continent, including mine workers recruited on a temporary basis from neighboring countries – as previously under *Apartheid* (Crush 2003). After 1994, unauthorized entry grew enormously. Africans from as far away as Ghana, Nigeria, and DR Congo flocked to the South African "El Dorado." Many brought with them qualifications and experience in medicine, education, administration, and business. Others joined the informal economy as hawkers, street food sellers, or petty traders. There are no accurate figures on irregular migrants, with wildly conflicting estimates. Over a million migrants have been deported since 1994 (Crush 2003). Most irregular migrants enter legally, although some risk injury crossing the border fence illegally. However, hostility to immigration – including outbreaks of violence – has hindered the development of legal mechanisms for employment and regularization.

Historically, African migration has been predominantly intracontinental. Today, despite media hysteria on the growth of African migration to Europe, actual numbers seem quite small (de Haas 2008). However, migration of Africans to Europe and the Middle East has increasingly involved irregular labor migrants. The areas receiving the most international attention include those closest to the Mediterranean coast – namely, Spain, Italy, and Malta. An estimated 31,000 illegal migrants arrived in the Canary Islands (part of Spain) in 2006, a 600% increase from 2005. At least 6,000 others were thought to have died in their attempts to reach the islands. This movement has since declined – partly as a result of EU financial support for preventive measures by African states. Africans who successfully reach Europe often face unemployment, racism, and homelessness (Castles and Miller 2009, pp. 156–158).

Asia-Pacific

This vast region, home to over half the world's population, has experienced many migratory flows over the last half-century, including refugees; postcolonial movements to European countries; migration of skilled workers to North America, Oceania, and Europe; labor deployment to the Gulf oil states; and, since the 1980s, labor movements from countries experiencing slower economic growth (such as the Philippines, Indonesia, Burma, and Bangladesh) to Japan and the "tiger economies" (Taiwan, South Korea, Singapore, Malaysia, and Thailand).

Many of the movements have been spontaneous and irregular. Even where migration appears to be highly regulated – for instance, in Japan and South Korea – the authorities tolerate "side doors" such as training schemes or the "back door" of irregular migration to meet labor needs. Similarly, the conditions of Asian migrant workers recruited for the Gulf oil state appear to be highly regulated, but this only applies to control of entries and residence, and not to protection of human rights. Workers are not allowed to settle or bring in dependants and are often segregated in barracks. Migrants can be deported for misconduct and often have to work very long hours. Many migrant workers are exploited by agents and brokers who take large fees (up to 25% of wages), but often fail to provide the jobs and conditions promised.

Box 9.4 Republic of Korea

In May 2011, 168,450 irregular migrants were recorded in the Republic of Korea (RoK), that is, 12.4% of the foreign population and 0.3% of the total population. Irregulars include overstaying low-skilled work visa holders, spouses of Korean nationals (mostly women), and tourists or short-term visa holders (Korea Immigration Service 2011; Statistics Korea 2011). Inflows of irregular migrants began to increase rapidly when the government eased entry rules to attract tourists to the Seoul Olympic Games in 1988 (Seol 2001). Although the RoK was facing labor shortages, the government did not recognize demand for low-skilled migrant labor. Instead, it tacitly condoned the employment of irregular migrants. The Industrial Trainee System of 1994 was the first official labor migration scheme. Ironically, the "industrial trainees" were paid less than irregular migrants but were not allowed to change their workplace. They became irregular by leaving their designated jobs or by overstaying (Seol 2001; Hur and Lee 2008). The number of irregular migrants peaked at 308,165 in 2002.

The government replaced the failed trainee scheme with the Employment Permit System (EPS) in 2005. Recruitment and placement is now handled by government agencies, aiming to provide transparent and low-cost processes. The government also introduced a scheme exclusively for members of Korean diasporas in 2002 (Lee 2010). The number of irregular migrants decreased considerably after the introduction of these policies (Korea Immigration Service 2011).

(continued)

Box 9.4 (continued)

The success of the EPS will depend on whether the first batch of workers move into irregularity when their contracts terminate in 2011 (Kim 2010).

Recruitment under the EPS is still costly (partly because of Korean language requirements) and highly competitive. This applies especially to female migrants who are less favored by manufacturing employers. In the absence of other opportunities, marriage migration may be an attractive option (Kim et al. 2007; Lee 2008). Brokers arrange many international marriages at short notice, and all costs are paid by the Korean spouses (Goh et al. 2005). But marriages do not always go smoothly because of differing expectations and difficulties in communication, and may end up by divorce. Marriage migrants cannot apply for visas or citizenship without the endorsement of their spouses. A divorced marriage migrant without citizenship loses her visa unless she provides clear evidence of domestic violence (Lee 2008). Some marriage migrants simply run away from abusive situations. Their spouses may then report them to the immigration office, which may cancel their visas, leaving the wives as irregular.

Immigration raids and deportation are still the primary enforcement strategies regardless of ongoing debates over their legality (Jung 2011). Officials are sometimes under pressure to meet allotted arrest quotas and resort to excessive use of force, which can lead to injuries or even death of irregular migrants. The leaders of migrant trade union are often targeted for deportation (Amnesty International 2011). The conditions of detention facilities have not significantly improved, even after a tragic fire of Yeosu Protection Centre in 2007, which killed ten inmates. Degrading treatment, overcrowding, and denial of medical access are issues of concern (Amnesty International 2009; Seol et al. 2005). The RoK government has also run assisted voluntary departure programs. Participation in such programs leads to exemption from fines and the prohibition of reentry, but without financial assistance (Korea Immigration Service 2010).

The employers of irregular migrants are rarely punished. In fact in 2009, all such employers were granted amnesty. The government explicitly stated that it eased the regulations in order to assist medium- and small-sized businesses and "to improve its image to be business friendly" (Korea Immigration Service 2009).

Irregular migration within Asia has grown rapidly since the 1980s. Thai workers move to Malaysia and other countries for work, while Thailand itself hosts up to 1.7 million undocumented workers, mainly from Burma (IOM 2005, pp. 110–112). Hugo points to the complexity of irregular migration, arguing that there is a continuum from voluntary individual movement, through use of middlemen, to trafficking and bonded labor (Hugo 2005, p. 25). The growth of irregular migration is linked to the unwillingness of governments to effectively manage migration and to the desire of employers for easily available and exploitable workers. Spontaneous undocumented migration can meet labor needs effectively but creates a situation of insecurity and "rightlessness" for workers. Moreover, they can become scapegoats for social problems like crime, disease, and unemployment.

Box 9.5 Malaysia

Malaysia was estimated in 2006 to have around one million undocumented workers, mainly from Indonesia, the Philippines, and Thailand, in its total migrant labor force of about 2.6 million (Skeldon 2006). The Malaysian Immigration Department has regularly carried out mass deportations of irregular migrants, particularly of Indonesians. In 2002, hundreds of thousands of irregular Indonesian workers were deported following a newly enacted law, which also provided for caning of irregular migrants. In 2003, 42,935 irregular immigrants were arrested under these laws, of whom around 9,000 were subjected to corporal punishment. Overall, 465,878 irregular migrants were arrested between 2000 and 2004. All irregular migrants – whether economic migrants or asylum seekers – face a sentence of up to 5 years and a fine of RM 10,000 (equal to US$2,600) (Kaur 2008, p. 14).

Particularly controversial has been the use of the RELA Corps (*Ikatan Relawan Rakyat Malaysia* or "Volunteers of the Malaysian People"). RELA is a civilian volunteer corps, first launched in 1972 to assist the government in maintaining order and security. Today, their main duty is to check the identity cards of foreigners including tourists and visitors. They are authorized to conduct interrogations at any time. RELA members have allegedly destroyed the ID cards of regular migrants in order to justify deportations (Kaur 2008). They have also been criticized for not distinguishing between undocumented migrants, refugees, and asylum seekers, even when UNHCR has issued cards certifying such status (Human Rights Watch 2007). In January 2011, the Malaysian government decided to increase the number of RELA members to nearly 2.6 million. Immediate concerns were raised in the Asian Forum for Human Rights and Development (Lee 2011).

Irregular Migration and Development

Migration has the potential to bring benefits for origin and destination countries and for migrants themselves (UNDP 2009). Human mobility is an integral part of human development as articulated by social philosopher and economist Amartya Sen (1999). For Sen (1999), development represents the "process of expanding the real freedoms that people enjoy" (p. 3). Mobility is a basic freedom and has the potential to lead to greater human capabilities. Reducing migration restrictions and ensuring that people can move safely and legally helps enhance human rights and can also lead to greater economic efficiency and social equality (UNDP 2009). This situates development as much more than economic growth in itself. However, as de Haas (2009) argues:

> restrictive immigration policies and marginalization of regular and irregular immigrants do not only involve considerable risk and suffering for the migrants involved, but are also likely to have a negative impact on the poverty and inequality-reducing potential of migration. (p. 55)

Irregular migration often results in high levels of risk and exploitation for migrants, which can seriously undermine development benefits. Also, in many cases, the movement of people is not the result of an increase in the opportunities and choices available to them. Migrant labor – in particular that of irregular migrants – is channeled into "three-D" jobs: dangerous, degrading, and dirty (Taran and Geronimi 2003). For many irregular migrants, and for all trafficked people, migrating actually represents a contraction of choices and freedoms (de Haas 2009).

According to current dominant views on migration and development, migration can bring benefits in various ways such as financial remittances, social remittances (transfer of development-friendly attitudes to origin areas by migrants), technology transfer, and "brain circulation." However, the development benefits of irregular migration may be much more limited than those of regular migration. While irregular migrants do send back remittances (they may even remit a higher proportion of their pay than regular migrants), their pay is likely to be lower, so the overall amount may be less. Also, irregular migrants may have to pay higher transfer rates than others, due to exclusion from official banking systems. Similarly, irregular migrants are less likely than regular migrants to bring back positive social remittances. Indeed, their experience of risky migration and living underground in an unwelcoming society may lead to negative reactions. Irregular migrants are also unlikely to play a big part in technology transfer. Informal sector employment in destination countries does not generally provide access to advanced technology or to training and education. Similarly, while "brain circulation" implies the gaining of new knowledge and techniques through work overseas, this applies mainly to regular skilled migrants and rarely to irregular ones.

Irregular migration is thus unlikely to have positive development benefits. If there is no willingness in an origin country to bring about development-friendly changes (such as improving governance and administration, tackling corruption, and developing infrastructure), then the benefits of migration will be limited to poverty reduction for specific families and increased consumer spending. Similarly, if there is no willingness in the country of destination to implement the reforms needed to maximize the benefits and minimize the risk of migration for irregular migrants, the positive development impact of irregular migration is likely to be limited to the contribution brought by economic remittances at the expense of migrants' basic rights and freedoms (Sabates-Wheeler 2009).

Improving Policy Responses to Irregular Migration

In its 2010 *World Migration Report*, IOM states that it "considers the curbing of irregular migration to be a priority" and suggests the building of effective capacities in ten specific areas (IOM 2010b, pp. 30–31). We cannot hope to cover all these topics in the space available to us, but, in any case, it is notable that many of the strategies suggested in the report – and indeed most measures adopted by states and intergovernmental organizations – are concerned with identifying, apprehending, detaining, and deporting irregular migrants (see Box 9.6 for examples). This can lead to violations of the human rights of migrants and can be counterproductive, as it forces migrants underground and makes them even more vulnerable to exploitation and abuse by smugglers and employers.

Box 9.6 Regional Strategies to Prevent and Control Irregular Migration

In 2004, the EU established the European Agency for the Management of Operational Cooperation at the External Borders of the Member States of the European Union (FRONTEX). In 2007, FRONTEX's role was expanded to include the establishment of the Rapid Border Intervention Teams (RABIT) for assistance to Member States in cases of mass influx situations. Border guards were to be trained for deployment in emergency cases. Amnesty International (AI) and the European Council on Refugees and Exiles (ECRE) published a joint statement stressing their concern that increased powers for the agency should be accompanied by enhanced assessment of compliance with fundamental rights of irregular migrants. AI and ECRE also noted their concerns that "allowing FRONTEX to implement technical assistance projects through funds from the financial instruments set up by ECHO and EuropeAid could lead to the use of humanitarian and development assistance for border control purposes" (Amnesty International and ECRE 2010, p. 7). Following an official request from Italy, FRONTEX launched Operation Hermes for the surveillance of the Mediterranean between Italy and North Africa. In this period, the no-fly zone with Libya came into effect and combat operations started in March 2011.

Similarly, memoranda of understanding (MoU) have been signed between North and Central American countries to enhance border control and reduce irregular migration. In particular, Mexico has signed MoUs with the USA and Central American countries for the "safe, orderly, dignified, and humane" repatriation of irregular migrants. In 2004, Mexico signed a MoU with the USA facilitating the deportation of an annual average of 570,000 Mexicans from the USA. In 2006, Mexico signed a similar MoU with Guatemala, Honduras, El Salvador, and Nicaragua. This scheme has in turn seen an annual average of 59,000 Central American irregular migrants deported from surveillance of the irregular migrant population.

(continued)

Box 9.6 (continued)

Regional strategies include the fight against irregular transit migration. In such cases, the countries of final destination put pressure on transit countries, and migrants become stranded before they reach their targeted destination. Countries on the borders of attractive destinations (e.g., Mexico, Turkey, and North African countries bordering the USA and the EU) become buffer areas for immigration control. For instance, the European Neighborhood Policy builds partnerships with transit countries: border control officers are offered special training, and readmission agreements are encouraged. However, transit countries can use such situations for their other ends: Turkey utilizes its role in transit migration as a negotiation tool in accession negotiations with the EU.

Even measures like "capacity-building in transit states" (IOM 2010b, p. 31) can be problematic, as they can involve the reinforcement of policing, border control, and detention capacities in states that do not adhere to human rights conventions. The engagement of Italy, Malta, and the UK with Libya is an example of this. In spite of a dire human rights record and evidence of abuse against asylum seekers and migrants, EU members actively sought the collaboration of Libya to control the flow of migrants and asylum seekers to Europe. In February 2003, the British government suggested setting up "transit processing centers" for asylum seekers in countries like Libya and Ukraine. The proposals were not implemented in this form, but they helped create a climate in which asylum seekers were seen as a security threat, justifying ever-tighter legal procedures and increased use of detention and deportation (Amnesty International 2010; Blitz 2007; Noll 2003).

Practical Measures to Address Irregular Migration

There is a strong case that irregular migration should be eliminated, because it undermines the rule of law, leads to violations of the human rights of migrants, and creates divisions and conflicts between migrants and significant groups in the populations of receiving countries. However, the main cause of irregular migration is not disregard of regulations by migrants, but rather the growth of inequality within and between countries, and the failure of states to create adequate migration regimes to meet economic demand. Strategies to respond to irregular migration need to address its fundamental causes, while safeguarding the rights and legitimate interests of all concerned. Here we discuss some recent practical responses to irregular migration. Some of these were discussed at the fourth meeting of the Global Forum on Migration and Development (GFMD) in Puerto Vallarta, Mexico, in 2010.

Assisted Voluntary Return and Reintegration Programs

IOM promotes Assisted Voluntary Return and Reintegration programs (AVRR) as an indispensable part of a comprehensive approach to migration management that combines efficient border management, effective asylum processing structures, and respect for human rights by facilitating the safe and dignified return of migrants and encouraging their sustainable reintegration in their country of return (IOM 2011). AVRR started with facilitating return transportation for asylum seekers from Germany in 1979, but later evolved into comprehensive projects promoting "the sustainability of returns" and preventing "irregular emigration at any cost." AVRR programs now also address the concerns of communities of return to "address the root causes of irregular migration." Target groups include "migrants who either have applied for asylum or have seen their asylum applications rejected, and other migrants currently residing or stranded in host countries who are willing to return voluntarily to their country of origin" (IOM 2011).[1]

A recent report (Black et al. 2011) argues the voluntary return programs are "non-coercive return programs" or "pay-to-go scheme" in the sense that irregular migrants choose voluntary return only when they are facing the prospect of forced removal. In spite of some potentially positive effects, such programs have never been very successful except when the program was supported by "the real threat of removal" like the Germany-Bosnia program of the late 1990s. However, the "non-coercive return programs" were still considered an important element in the policy mix to reduce irregular migration.

UK studies show that the "non-voluntary" character of AVRR programs is even more problematic when these involve the return of asylum seekers. Webber (2011) found that most of the return projects were implemented without assured safety of returned asylum seekers, sufficient assistance for sustainable return, and fully voluntary decisions.

---

[1]From 2000 to 2010, AVRR was implemented in 45 countries to return over 400,000 irregular migrants to 174 countries. Between 2008 and 2010, the UK, Germany, Austria, the Netherlands, and Belgium returned the most irregular migrants; Brazil, Iraq, the Russian Federation, Kosovo (Serbia), Serbia, and Mongolia were the main destinations (IOM 2010a, b). The AVRR operates in three stages. The predeparture stage includes return counseling, provision of origin country information, travel allowances, and reinstallation grants; the transportation stage includes transport and medical checks; and the postarrival stage includes reception, health-related support, provision of reintegration grants, support for vocational training, and income-generating activities.

Blitz et al. (2005) suggest that return migration programs were driven by "domestic interest-based arguments" and used as a means of "placating anti-immigrant public opinions." A study of France and Spain also shows that voluntary return policies were adopted in the context of the economic crisis but were not successful in alleviating the effects of the crisis (Plewa 2009).

Regional Consultative Processes (RCPs)

Although some RCPs discuss a broad range of migration topics, most of them focus on specific issues such as reducing irregular migration (e.g., the Budapest Process in Europe) or labor migration (e.g., the Colombo Process, composed of Asian sending countries). RCPs generally aim at building trust and understanding migration better through breaking down national and departmental divides, creating networks, harmonizing positions, and building capacities.

Combating smuggling, trafficking, and irregular migration is a key issue for most RCPs. For example, the Intergovernmental Consultations on Migration, Asylum and Refugees (IGC), established by 17 major Western destination countries and UNHCR, IOM, and the European Commission in 1985, dealt with the issues including "return, smuggling and trafficking,…illegal migration, burden sharing, and the use of technology in the management of migration" (IGC 2011). The Bali Process on People Smuggling, Trafficking in Persons and Related Transnational Crime, with some 43 member countries from the Asia-Pacific region and 30 observers from other parts of the world and international organizations, aims at effective information and intelligence sharing; cooperation on border and visa systems; cooperation on verifying the identity of irregular migrants; and tackling the root causes of irregular migration by increasing opportunities for legal migration (Bali Process 2011).

Governments are most likely to engage actively in RCPs when migration plays a big part in their domestic politics. The Bali Process was established with support from the Australian government immediately after the Tampa Affair of 2001 (when a Norwegian ship that had rescued 438 Afghan asylum seekers from a sinking vessel was refused permission to land them in Australia). The first Ministerial Conference of the Bali Process in 2002 emphasized cooperation to combat "criminal activities." The conference also noted "possible links between terrorist elements and people smuggling and trafficking operation" (Bali Ministerial Conference 2002). The Australian government's 2011 agreement with Malaysia to transfer asylum seekers from Australia to Malaysia in return for resettlement of recognized refugees in the opposite direction was partly a result of bilateral discussions during a ministerial meeting of the Bali Process (Bowen 2011). This plan was ruled unlawful by the High Court of Australia in August 2011 (Millar 2011), but, at the time of writing, the Australian government is seeking to change the Migration Law to allow the deal to go ahead.

Circular or Temporary Migration Schemes

Circular or temporary migration schemes are increasingly advocated as a way of reducing irregular migration. One of the most commonly cited examples is to be found in Spain. According to the agreement on a pilot project, Moroccan workers were to be employed through a circular migration program to work on strawberry plantations. Initially, however, the return of the workers to Morocco at the end of their contract was so low that the authorities changed the rules so that only mothers under the age of 40 with dependent children could participate. Such circular or temporary migration schemes violate human rights, as they include discriminatory selection criteria. Such schemes may be better than irregular migration, "yet the comparison should be with the conditions of national workers, not workers in irregular status" (Wickramasekara 2011, p. 48).

An argument often used in favor of temporary migration schemes is that of a "trade-off between the numbers and the rights" of low-skilled migrants (Ruhs and Martin 2008). The economic logic behind this argument is that more rights for migrants implies higher labor costs, which lowers demand for migrant labor. In other words, existence of more rights results in fewer migrants being employed and vice versa. This argument follows the experience of immigration to the authoritarian regimes of the Persian Gulf (Cummins and Rodríguez 2010). While these countries indeed have very high levels of immigration, low-skilled migrants work under exploitative or abusive conditions below international standards laid out in ILO and UN conventions (Human Rights Watch 2010a). To follow the lead of these countries in terms of immigration policy would be to support a race to the bottom which creates and legitimizes second-class citizenship for low-skilled migrants.

The work of trade unions, migrant associations, and NGOs is essential for the protection of the human rights of workers in circular and temporary migration schemes (Avendaño 2009, pp. 13–16). The ILO's Bureau for Workers' Activities (ACTRAV) has initiated agreements between workers' organizations in origin and destination countries. Signatories to the trade union agreements also acknowledge their commitment to promote the ratification of and respect for ILO Migrant Workers Conventions and to actively fight against racism and discrimination. In May 2009, the first three model bilateral cooperation agreements for the protection of human rights of migrant workers were signed between trade unions in Sri Lanka and Bahrain, Jordan, and Kuwait. In 2009, the International Trade Union Confederation for the Americas (TUCA) also agreed to promote the model agreement in their region (Wickramasekara 2011, p. 82).

Respect for Human Rights of All Migrants

International law does not differentiate between regular and irregular migrants with regard to entitlement to basic human rights. The 1990 International Convention on the Protection of the Rights of All Migrant Workers and Members of Their Families

(ICRMW) confirms that all migrants are entitled to basic human rights such as the right to life, right not to be in slavery or servitude, freedom of thought and speech, right to fair legal treatment, and economic, social, and cultural rights. Although the effectiveness of the ICMRW is often questioned due to the low number of ratification (45 state parties) and lack of participation of countries of destination, the core provisions of human rights of all migrant workers are already endorsed broadly as human rights of all human beings through other international instruments like the International Covenant on Economic, Social and Cultural Rights and the International Covenant on Civil and Political Rights.

The Global Migration Group (GMG), which is composed of 14 major international organizations, reconfirmed that the fundamental rights of all persons apply to all migrants regardless of their migrant status (Global Migration Group 2010). The Group also emphasized:

> [a]lthough States have legitimate interests in securing their borders and exercising immigration controls, such concerns cannot, and indeed, as a matter of international law do not, trump the obligations of the State to respect the internationally guaranteed rights of all persons. (Statement of the GMG on the Human Rights of Migrants in Irregular Situation, para. 4)

Governments therefore need to ensure respect for the human rights of migrants, regardless of their immigration status. A key issue here is access to health and education. Currently, some regional governments in Spain and Italy allow access to health and education services for irregular migrants and their children. This is partly due to the refusal of education and health workers to discriminate. In other countries, like Britain and Germany, governments require educational and medical personnel to report irregular migrants to them, but many professionals refuse to do so.

Civil Society Organizations

The work of civil society organizations is important in exposing and combating discrimination, as well as in drawing public attention to the positive contribution of regular and irregular migrants to host societies. In Europe, the best-known nongovernmental organization (NGO) is the Brussels-based Platform for International Cooperation on Undocumented Migrants (PICUM). PICUM produces reports and newsletters and works in capacity-building for NGOs and other actors involved in preventing and addressing discrimination against irregular migrants. They carry out awareness raising and advocacy to include irregular migrants in social and integration policies on the national and European levels. They also contribute to international dialogue on migration within UN agencies and other organizations (See PICUM 2011).

Other umbrella organizations include ECRE, consisting of 76 refugee-assisting agencies in 30 countries; the European Network Against Racism (ENAR), consisting of NGOs working to combat racism in all the EU countries; and the campaign for the adoption of the UN Migrant Convention: December 18. Another example at the regional level is the Platform for European Red Cross Cooperation on Refugees, Asylum-Seekers and Migrants (PERCO). PERCO aims to develop and strengthen Red Cross activities for and with refugees and migrants on a national and international level and promote cooperation between the European National Red Cross and Red Crescent Societies. Church organizations such as Caritas Europa, the Churches' Commission for Migrants in Europe (CCME), the International Catholic Migration Commission (ICMC), the Jesuit Refugee Service Europe (JRS Europe), and the Quaker Council for European Affairs (QCEA) also provide shelter and seek to meet the urgent needs of irregular migrants. The HUMA network project was a network initiated by Médecins du Monde. It consisted of 16 partners in 16 EU countries, promoting access to health care for irregular migrants and asylum seekers. The project ended in March 2011, but Médecins du Monde continues its support for health programs directed at irregular migrants.

NGOs and civil society are also taking an increasingly active role in North and Central America. Migrant organizations have become more vocal in the fight for the rights of irregular migrants in the USA. An example of this were the 2006 protests that saw thousands of people, including irregular migrants, rallying against proposed changes to US immigration policy that would have raised penalties for irregular immigration. In Mexico, NGOs have also taken a more active role. In addition to the advocacy work of NGOs such as Sin Fronteras I.A.P (Without Borders) and the Centro de los Derechos del Migrante (Center for Migrants Rights), many Casas del Migrante (Migrant Houses) have been established all over the country to provide social and legal assistance to irregularly migrants.

Prosecuting Criminal Organizations Involved in Trafficking and Smuggling

Governments generally acknowledge that criminal organizations involved in trafficking and smuggling of people are of a transnational character and are inclined to make joint agreements in this area. The countries of the Greater Mekong Subregion (Cambodia, People's Republic of China, Lao PDR, Myanmar, Thailand, and Vietnam) for instance have joined together in their efforts to stem the trafficking of people across their borders. In July 2010, the Global Plan of Action against Trafficking in Persons was endorsed by the UN General Assembly. NGOs are also active in assisting trafficked persons and organizing public awareness programs to highlight the significance of the issue. In Europe, KOK in Germany, STOP NOW in Greece, On the Road in Italy, and STV in the Netherlands are a few examples of NGOs, which have had long-term experiences in this field.

The GFMD Process

The Global Forum on Migration and Development (GFMD) was established to advance understanding and cooperation among countries on the relationship between migration and development and to foster practical outcomes. Up to 2010, the GFMD had held four major international meetings. Irregular migration was covered in specific roundtables in two of these meetings, and recommendations were made. However, it appears that there has been little systematic monitoring of the implementation of these recommendations; perhaps this should be an item at the beginning of each forum.

The first forum where the issue was covered was in 2008 in Manila. Roundtable 2.2 addressed "Managing Migration and Minimizing the Negative Impacts of Irregular Migration" and suggested that irregular status of migrants increased their vulnerability toward criminalization. "To conduct targeted research on costs-benefits and impacts of regular and irregular migration on development for developing countries with significant inward and outward migration flows" was recommended, yet this has not been implemented. The third meeting of the GFMD in Athens in 2009 did not include a roundtable on irregular migration but discussed criminalization of irregular migrants within other roundtables.

The fourth GFMD meeting in Puerto Vallarta in 2010 devoted two roundtable sessions to the issue, titled "Partnerships for more regular and protected migration" and "Joint strategies to address irregular migration." The latter discussed ways of developing partnerships and joint strategies, suggesting that irregular migration should continue to be on the agenda of the GFMD and other international conferences. Joint strategies and partnerships were also dealt with in the Civil Society-Government symposium held in Geneva in August 2011 in the lead-up to the GFMD Concluding Debate.

Two thematic meetings on irregular migration were held in El Salvador and Istanbul in October 2011. The specific objectives of both meetings were to exchange information on bilateral, regional, and international processes; to protect irregular migrants; and to identify mechanisms for effective partnerships. The meetings "confirmed the observation of the UNDP Human Development Report, 2009, that policies and programs to ensure that people move safely and legally can also result in greater economic efficiency, social equality and contribution to development" (GFMD 2011, p. 43). Countries of origin were urged to strengthen the use of consular networks abroad; origin and destination countries were admonished to establish national councils, committees, or working groups for the support, protection, and human development of migrants and their families; and the GFMD was challenged to establish a working group to test and operationalize the models offered by protection-oriented instruments such as the ILO Multilateral Framework on Labour Migration and similar tools from expert international organizations.

These outcomes of the thematic meetings in 2011 were taken forward to the GFMD Concluding Debate, where governments hoped to explore practical examples of regularization options and partnership programs. The thematic meetings demonstrated that while there were numerous bilateral and regional initiatives and dialogues to address irregular migration and ways to mitigate migration pressures through development cooperation, the migration policies of developed countries continue to be defined independently of their impacts on developing economies. The challenge remains profound, and it remains to be seen whether the GFMD will be able to make game-changing contribution in the years ahead.

Conclusions

Irregular migration seems to be an almost universal phenomenon, although it has only been seen as a matter for concern in relatively recent times. Where governments fail to create legal migration systems despite strong employer demand for workers, migrants experience high levels of risk and exploitation. Smuggling, trafficking, bonded labor, and lack of human and worker rights are the fate of millions. Existing policy approaches have often been lacking in vision and are not always based on an adequate understanding of the drivers of migration. Simplistic policy solutions often have perverse effects. For example, one effect of increased control on the US-Mexico border is that temporary labor migrants have decided to settle permanently in the USA to avoid the risks of the journey.

Together with other cross-border flows – of commodities, capital, intellectual property, and culture – human mobility is an integral part of globalization. At a time when regulatory frameworks for finance, trade, and many other aspects of international cooperation have become important, global governance of migration remains weak. Most of the world's nations – and especially the labor importers – have refused to sign the 1990 Convention on the Rights of Migrant Workers and Members of Their Families, as well as earlier ILO Conventions. There is no global body to manage cooperation and standard-setting on international migration, in the same way as the World Bank and the International Monetary Fund (IMF) for finance, or the World Trade Organization (WTO) for trade.

Irregular migration has become a highly political issue. All too often the underlying approach seems to be one of "muddling through," in which politicians and officials seem to tacitly believe that some irregular migration is inevitable, and that current policies can provide acceptable compromises to meet the needs of both employers and policy makers. This is effectively what is happening in most destination countries. In public discourse, on the other hand, many political leaders call for stricter border control and for the complete elimination of irregular migration. This line plays well with the media but is rarely carried out in practice, because it would have very high implementation costs and could lead to serious problems of unmet labor demand.

In discussions within regional bodies like the European Union, ASEAN, or MERCOSUR, there is more openness to rational planning. Such bodies tend to advocate programs of migration management based on such measures as temporary and circular migration; restrictions on migrants' labor rights (e.g., the rights to change jobs);

regularization programs to integrate existing irregular migrants; and strict entry rules to prevent future irregular migration. Effectively, such programs often mean managing migration to meet the economic and political objectives of destination countries.

An alternative approach would be to create systems of legal migration based on recognition of the interests of all stakeholders and the human rights of all migrants. This would mean setting up cooperative arrangements (or institutions), which would give a voice not only to stakeholder groups in destination countries (governments, employers, trade unions, and affected communities) but also to similar groups in origin countries and to the migrants themselves. The key principle would be the recognition of the human rights of migrants, including the right to move across borders in safety and dignity and to participate in political, economic, social, and cultural affairs. Modes of participation would vary: permanent settlers should quickly gain secure residence rights and a pathway to citizenship, while temporary migrants would be more interested in fair access to labor markets, in health care and social protection, and in the portability of earned entitlements when they leave. A precondition for this approach would be the fair assessment of labor needs in destination countries and the enforcement of the principle of equal pay and conditions for equal work.

In our view, this cooperative approach offers the best prospect for solutions that benefit migrants and their communities of origin, as well as destination countries. An important first step would be to highlight the reality of irregular migration in order to counter stereotypes that link migration to crime, insecurity, and terrorism. As stated during the 2010 Global Forum on Migration and Development:

> Migrant involvement in key growth sectors such as agriculture, construction and services, contributes significantly to the welfare of the population. For example, domestic work and care of children and the elderly, usually poorly paid and stigmatized, is a fundamental contribution of migrants to the welfare of families at both ends of the migration spectrum. (GFMD 2010, p. 5).

A fundamental change in attitudes would be an important step toward fairer and more effective migration policies. Migrants leave their countries of origin in order to improve their livelihoods and to gain greater security. Migration should be seen not as a threat to state security, but as a result of the human insecurity that arises through global inequality. Migration is thus an important aspect of human development as articulated in Amartya Sen's principle of "development as freedom" (Sen 1999) (see above). This perspective could provide a new frame of reference when thinking about migrant rights, as well as about state policies and public attitudes.

This is a time of potential change. The 2007–2010 global economic crisis has disrupted patterns of migration and brought about declines in remittances to poor communities that had become dependent on them. However, migration and migrant populations have proved more resilient than many expected. Developed countries and newly industrializing countries are competing for scarce skills, while even the supply of lower-skilled labor is beginning to appear finite in view of the demographic transition taking place throughout the world. The labor-importing countries – after many years of refusal – have begun to talk to governments of migrant-origin countries through such mechanisms as the UN High Level Dialogue on Migration and Development in 2006, and the Global Forum on Migration and Development.

In the past, some analysts have argued that facilitating development in migrants' countries of origin might reduce international migration. On this basis, governments have seen development policies as a way of addressing the "root causes" of migration (Castles and Van Hear 2011). However, more recent migration research has shown that development actually increases migration, by giving people the resources to move (Martin and Taylor 2001).

These considerations lead to the uncomfortable conclusion that irregular migration is likely to be with us as long as there is marked global inequality. No amount of border regulation or return policies will stop people moving, when the potential rewards are so high. Human rights advocates therefore need to pursue the dual agenda of working for long-term reform of the global economic order, while in the short-term working for the respect of the rights of all migrants, whether regular or irregular.

References

Abella, M. (2005). Clandestine migration. In M. J. Gibney & R. Hansen (Eds.), *Immigration and asylum: From 1900 to the present* (Vol. 1). Santa Barbara: ABC-CLIO.

Alba, F. (2010). *Mexico: A crucial crossroads*. Retrieved July 8, 2011, from http://www.migrationinformation.org/feature/display.cfm?ID=772

Amnesty International. (2009). *South Korea: Disposable labor: Rights of migrant workers in South Korea*. London: Amnesty International.

Amnesty International. (2010). *Libya of tomorrow: What hopes for human rights*. London: Amnesty International.

Amnesty International. (2011). South Korean government must not deport migrants' trade union leader. In *Amnesty International public statement*. Seoul: Amnesty International.

Amnesty International, & ECRE. (2010). *Briefing on the commission proposal for a regulation amending council regulation (EC) 2007/2004 establishing a European Agency for the Management of Operational Cooperation at the external borders of the member states of the European Union (FRONTEX)*. Brussels: Amnesty International/ECRE.

Amnistía Internacional. (2010). *Víctimas invisibles: migrantes en movimiento en México*. Retrieved July 8, 2011, from http://www.amnesty.org/es/news-and-updates/abusos-generalizados-migrantes-mexico-crisis-ddhh-2010–04–28

Anderson, B. (2000). *Doing the dirty work: The global politics of domestic labor*. London: Zed Books.

Avendaño, A. (2009). *Reintegration and circular migration – Effective for development?* Paper presented at the Session 2.2 Session 2.1, Roundtable 2: Migrant integration, reintegration and circulation for development, Civil Society Days, Global Forum on Migration and Development. Athens: Global Forum on Migration and Development.

Bakewell, O., & de Haas, H. (2007). African migrations: Continuities, discontinuities and recent transformations. In P. Chabal, U. Engel, & L. de Haan (Eds.), *African alternatives* (pp. 95–117). Leiden: Brill.

Bali Ministerial Conference. (2002, February 26–28). *Co-Chair's statement*. Paper presented at the Bali Ministerial Conference on People Smuggling, Trafficking in Persons and Related Transnational Crime, Bali, Indonesia.

Bali Process. (2011). *Welcome to Bali Process*. Retrieved July 26, 2011, from http://www.baliprocess.net

Bélanger, D., Lee, H.-K., & Wang, H.-Z. (2010). Ethnic diversity and statistics in East Asia: 'Foreign brides' surveys in Taiwan and South Korea. *Ethnic and Racial Studies*. doi:10.1080/01419870903427507.

Berggren, E., Likic-Brboric, B., Torsöz, G., & Trimiklionitis, N. (Eds.). (2007). *Irregular migration, informal labor and community: A challenge for Europe*. Maastricht: Shaker Publishing.
Bhagwati, J. (2003). Borders beyond control. *Foreign Affairs, 82*(1), 98–104.
Bigo, D., & Guild, E. (Eds.). (2005). *Controlling frontiers: Free movement into and within Europe*. Aldershot: Ashgate.
Black, R., Collyer, M., & Somerville, W. (2011). Pay-to-go schemes and other noncoercive return programs: Is scale possible? In *Improving US and EU immigration systems*. Washington: Migration Policy Institute.
Blitz, B. (2007). *Libyan nationals in the United Kingdom: Geo-political considerations and trends in asylum and return* (Vol. Working Paper). Oxford: COMPAS.
Blitz, B. K., Sales, R., & Marzano, L. (2005). Non-voluntary return? The politics of return to Afghanistan. *Political Studies, 53*(1), 182–200. doi:10.1111/j.1467-9248.2005.00523.x.
Bloch, A. (2003). Victims of trafficking or entrepreneurial women? Narratives of post-soviet entertainers in Turkey. *Canadian Woman Studies, 22*(3–4), 152–158.
Bowen, C. (2011). *Joint statements by the Prime Ministers of Australia and Malaysia on a regional cooperation framework*. Retrieved June 27, 2011, from http://www.minister.immi.gov.au/media/cb/2011/cb165099.htm
Castles, S., & Miller, M. J. (2009). *The age of migration: International population movements in the modern world* (4th ed.). Basingstoke/New York: Palgrave-Macmillan/Guilford.
Castles, S., & Van Hear, N. (2011). Root causes. In A. Betts (Ed.), *Global migration governance* (pp. 287–306). Oxford: Oxford University Press.
CEC. (2005). *Green paper on an EU approach to managing economic migration*. Brussels: Commission of the European Communities.
Clandestino. (2009a). *Comparative policy brief – Size of irregular population*. Hamburg: Clandestino Research Project.
Clandestino. (2009b). *Policy brief – Irregular migration in Italy: Counting the uncountable*. London: Clandestine Research Project.
Comisión Nacional de los Derechos Humanos. (2011). *Informe especial sobre secuestro de migrantes en México*. Retrieved July 6, 2011, from http://www.cndh.org.mx/#
Cornelius, W. A. (2001). Death at the border: Efficacy and unintended consequences of US immigration control policy. *Population and Development Review, 27*(4), 661–685.
Council of Europe. (2011). *The large-scale arrival of irregular migrants, asylum seekers and refugees on Europe's southern shores*. Strasbourg: Council of Europe, Committee on Migration, Refugees and Population.
Crush, J. (2003). *South Africa: New nation, new migration policy?* Retrieved January 25, 2008, from http://www.migrationinformation.org/
Cummins, M., & Rodríguez, F. (2010). Is there a number versus rights trade-off in immigration policy? What the data say. *Journal of Human Development and Capabilities, 11*(2), 281–303.
de Haas, H. (2008). *Irregular migration from West Africa to the Maghreb and the European Union: An overview of recent trends*. Geneva: International Organization for Migration.
de Haas, H. (2009). Research paper 1: Mobility and human development. In *Human development report research series*. New York: United Nations Development Programme.
Délano, A. (2009). From limited to active engagement: Mexico's emigration policies from a foreign policy perspective (2000–2006). *International Migration Review, 43*(4), 764–814.
Delgado Wise, R. (2007). The reshaping of Mexican labor exports under NAFTA: Paradoxes and challenges. *International Migration Review, 41*(3).
DIAC. (2009). *Initiatives to combat illegal work in Australia*. Canberra: Department of Immigration and Citizenship.
Düvell, F. (2006). Introduction and background. In F. Düvell (Ed.), *Illegal immigration in Europe: Beyond control?* (pp. 3–39). Basingstoke: Palgrave Macmillan.
Düvell, F., & Vollmer, B. (2009). Irregular migration in and from the neighborhood of the EU. A comparison of Morocco, Turkey and Ukraine. In *CLANDESTINO: Undocumented migration counting the uncountable. Data and trends across Europe*. Oxford: COMPAS.

Eder, M. (2007). Rethinking the state through the Moldavian new migrants in Turkey (in Turkish). *Toplum ve Bilim, 108*, 129–143.

Erder, S., & Kaska, S. (2003). *Irregular migration and trafficking in women: The case of Turkey*. Geneva: IOM.

Gargis, P. (2011, June 10). Alabama Governor sings nation's toughest immigrant law. *Reuters*.

GCIM. (2005). *Migration in an interconnected world: New directions for action: Report of the Global Commission on International Migration*. Geneva: Global Commission on International Migration.

GFMD. (2010). *Joint strategies to address irregular migration: Background paper*. Retrieved from http://www.gfmd.org/en/documents-library/docs-irregular-migration.html

GFMD. (2011, December 1–2). *Cluster II summary report for the working session on addressing irregular migration through coherent migration and development strategies*. At the Concluding Debate, Geneva.

Global Detention Project. (2009). *Italy detention country profile*. Retrieved July 15, 2011, from http://www.globaldetentionproject.org/countries/europe/italy/introduction.html

Global Migration Group. (2010). *Statement of the Global Migration Group on the human rights of migrants in irregular situation*. Retrieved September 11, 2011, from http://www.ohchr.org/EN/NewsEvents/Pages/DisplayNews.aspx?NewsID=10396&LangID=E

Goh, H. U., Kim, H. M., Soh, R., Kim, J. S., & Kim, J. (2005). *International marriage broker system: Survey report in Vietnam and the Philippines*. Presidential Committee on Social Inclusion (in Korean).

Guild, E. (2009). *Security and migration in the 21st century*. Cambridge: Polity.

Harris, N. (2002). *Thinking the unthinkable: The immigration myth exposed*. London: I.B. Tauris.

Honorable Congreso de la Unión. (2011). *Ley de Migración*. Retrieved July 6, 2011, from http://www.diputados.gob.mx/LeyesBiblio/index.htm

Hugo, G. (2005). *Migration in the Asia-Pacific region*. Geneva: Global Commission on International Migration.

Human Rights Watch. (2007). Malaysia: Disband abusive volunteer corps: Government uses RELA force on migrant workers. Retrieved July 15, 2011, from http://www.hrw.org/en/news/2007/05/08/malaysia-disband-abusive-volunteer-corps

Human Rights Watch. (2010a). *Rights on the line: Human rights watch work on abuses against migrants in 2010*. New York: Human Rights Watch.

Human Rights Watch. (2010b). *"Tough, fair, and practical" – A human rights framework for immigration reform in the United States*. New York: Human Rights Watch.

Human Rights Watch. (2011). *World report*. New York: Human Rights Watch.

Hur, J.-J., & Lee, K. (2008, March 25–26). *Demographic change and international labor mobility in Korea*. Paper presented at the PECC-ABAC conference on "Demographic Change and International Labor Mobility in the Asia Pacific Region: Implications for Business and Cooperation", Seoul, Korea.

Icduygu, A. (2011). *The irregular migration corridor between the EU and Turkey: Is it possible to block it with a readmission agreement?* (Vol. EU-US Immigration Systems 2011/14). Florence: CARIM, European University Institute.

IGC. (2011). *Intergovernmental consultations on migration, asylum and refugees*. Retrieved September 8, 2011, from http://www.igc.ch/

ILO. (2006). *Realizing decent work in Asia: Fourteenth Asian regional meeting: Report of the Director-General*. Geneva: International Labor Office.

Instituto Nacional de Migración. (2010). *Boletín mensual de estadísticas migratorias 2010*. México D.F: SEGOB.

IOM. (2005). *World migration 2005: Costs and benefits of international migration*. Geneva: International Organization for Migration.

IOM. (2008). *World migration 2008: Managing global mobility in the evolving global economy*. Geneva: International Organization for Migration.

IOM. (2010a). *Irregular migration.* Retrieved June 25, 2010, from http://www.iom.int/jahia/Jahia/about-migration/managing-migration--irregular-migration

IOM. (2010b). *World migration report 2010.* Geneva: International Organization for Migration.

IOM. (2011). *Assisted voluntary return and reintegration: Annual report of activities 2010.* Geneva: International Organization for Migration.

Jimenez, M. (2009). *Humanitarian crisis: Migrant deaths at the U.S.-Mexico Border*. San Diego: American Civil Liberties Union of San Diego and Imperial Counties & Comisión Nacional de los Derechos Humanos.

Jones, S. (2000). *Making money off migrants: The Indonesian exodus to Malaysia.* Hong Kong/Wollongong: Asia 2000 and Centre for Asia Pacific Social Transformation Studies.

Jordan, B., & Düvell, F. (2002). *Irregular migration: The dilemmas of transnational mobility.* Cheltenham/Northampton: Edward Elgar.

Jung, J.-H. (2011). Human rights of foreigners. In K. Chung (Ed.), *Understanding Korean migration policy* (Immigration policy series, Vol. 1). Seoul: IOM Migration Research and Training Centre (in Korean).

Kaur, A. (2008). *International migration and governance in Malaysia: Policy and performance* (UNEAC Asia Papers No. 22). Armidale: University of New England Asia Pacific Centre.

Kim, G. (2010). 2010 Migration trend in Korea: Migrant workers. In *Global migration trends* (Vol. 2010–Vol. 3, Special ed.). Goyang: IOM Migration Research and Training Centre (in Korean).

Kim, H. M., Kim, G.-D., Kim, M.-J., Kim, J. S., & Kim, C. (2007). Research on labor and marriage migration process from Mongolia and Vietnam to Korea and the Impact on Migrant Rights. In *2007 joint project of inter-Asian NIs on current human rights issues.* Seoul: National Human Rights Commission (in Korean).

Korea Immigration Service. (2009). *Ministry of Justice, great easing of the regulation in visa issuance of employers who illegally hired foreigners, in order to support the lack of work force in medium and small sized business.* Retrieved July 13, 2011, from http://www.immigration.go.kr/ (in Korean).

Korea Immigration Service. (2010). *Illegally staying foreigners, "Assisted departure programme"*. Retrieved July 13, 2011, from http://www.moj.go.kr/ (in Korean).

Korea Immigration Service. (2011, May). *Monthly report on the statistics of immigration and foreigner policy.* Korea Immigration Service, Ministry of Justice (in Korean).

Lee, H.-K. (2008). International marriage and the state in South Korea: Focusing on governmental policy. *Citizenship Studies, 12*(1), 107–123. doi:10.1080/13621020701794240.

Lee, H.-K. (2010). Preference for co-ethnic groups in Korean immigration policy: A case of ethnic nationalism? *Korea Observer, 41*(4), 559–591.

Lee, T. (2011). *Oral statement delivered on behalf of Asian Forum for Human Rights and Development.* Forum Asia.

Lutz, H. (Ed.). (2008). *Migration and domestic work. A European perspective on a global theme.* Aldershot: Ashgate.

Lutz, H., Phoenix, A., & Yuval-Davis, N. (Eds.). (1995). *Crossfires: Nationalism, racism and gender in Europe.* London: Pluto Press.

Martin, P. L. (2004). The United States: The continuing immigration debate. In W. A. Cornelius, T. Tsuda, P. L. Martin, & J. F. Hollifield (Eds.), *Controlling immigration: A global perspective* (pp. 50–85). Stanford: Stanford University Press.

Martin, P. L., & Taylor, J. E. (2001). Managing migration: The role of economic policies. In A. R. Zolberg & P. M. Benda (Eds.), *Global migrants, global refugees: Problems and solution* (pp. 95–120). New York/Oxford: Berghahn.

Martin, P., Abella, M., & Kuptsch, C. (2006). *Managing labor migration in the twenty-first century* (Global management series). New Haven/London: Yale University Press.

Mazza, J., & Sohnen, E. (2010). *On the other side of the fence: Changing dynamics of migration in the Americas.* Retrieved June 2, 2010, from http://www.migrationinformation.org/Feature/display.cfm?id=784

McCabe, K., & Meissner, D. (2010). *Immigration the United States: Recession affects flows, prospects for reform*. Retrieved from http://www.migrationinformation.org/usfocus/display.cfm?ID=766

Migration Dialogue. (2011). *Migration news: Elections, DREAM, SB 1070, states*. Retrieved July 6, 2011, from http://migration.ucdavis.edu/mn/more.php?id=3648_0_2_0

Milanovic, B. (2007). Globalization and inequality. In D. Held & A. Kaya (Eds.), *Global inequality: Patterns and explanations* (pp. 26–49). Cambridge/Malden: Polity.

Millar, P. (2011, August 31). Court sinks Malaysia asylum deal. *The Sydney Morning Herald*.

Mingione, E., & Qassoli, F. (2000). The participation of immigrants in the underground economy in Italy. In R. King, G. Lazaridis, & C. Tsardanidis (Eds.), *Eldorado or fortress? Migration in southern Europe* (pp. 27–56). Basingstoke: Macmillan.

Network Migration in Europe. (2010). *Italy's migration experiences*. Retrieved July 15, 2011, from http://www.migrationeducation.org

Noll, G. (2003). *Visions of the exceptional: Legal and theoretical issues raised by transit processing centers and protection zones* (University of Lund Working Paper). Lund: University of Lund.

Ozcurumez, S., & Senses, N. (2011). Europeanization and Turkey: Studying irregular migration policy. *Journal of Balkan and Near Eastern Studies, 13*(2), 233–248.

Passel, J. S., & Cohn, D. V. (2009). *A portrait of unauthorized immigrants in the United States*. Washington, DC: Pew Hispanic Center.

Passel, J. S., & Cohn, D. V. (2011). *Unauthorized immigrant population: National and state trends 2010*. Washington, DC: Pew Hispanic Center.

Pastore, F. (2008). Modes of migration regulation and control in Italy. In J. Doomernik & M. Jandl (Eds.), *Modes of migration regulation and control in Europe* (pp. 105–128). Amsterdam: Amsterdam University Press.

Pécoud, A., & de Guchteneire, P. (Eds.). (2007). *Migration without borders: Essays on the free movement of people*. Paris: UNESCO.

PICUM. (2010). *PICUM's main concerns about the fundamental rights of undocumented migrants in Europe*. Brussels: Platform for International Cooperation on Undocumented Migrants.

PICUM. (2011). *Mission*. Available at http://picum.org/en/about-us/mission/

Plewa, P. (2009). *Voluntary return programmes: Could they assuage the effects of the economic crisis?* (Working Paper No. 75, Vol. WP-10–75). Oxford: Centre on Migration, Policy and Society, University of Oxford.

Portes, A. (1999). Conclusion: Towards a new world – The origins and effects of transnational activities. *Ethnic and Racial Studies, 22*(2), 463–477.

Republic of Turkey Ministry of Foreign Affairs. (2011). *Turkey's fight against illegal migration*. Retrieved July 15, 2011, from http://www.mfa.gov.tr/turkey-_s-fight-against-illegal-migration.en.mfa

Reyneri, E. (2001). *Migrants' involvement in irregular employment in the Mediterranean countries of the European Union*. Geneva: International Labor Organization.

Ruhs, M., & Anderson, B. (2006). *The origins and functions of illegality in migrant labor markets: An analysis of migrants, employers and the state in the UK*. Oxford: Centre on Migration, Policy and Society.

Ruhs, M., & Martin, P. (2008). Numbers vs. rights: Trade-offs and guest workers programs. *International Migration Review, 42*(1), 249–265.

Ruiz, I., & Vargas-Silva, C. (2009). Another consequence of the economic crisis: A decrease in migrants' remittances. *Applied Financial Economics, 20*(1), 171–182.

Sabates-Wheeler, R. (2009). *Research paper 26: The impact of irregular status on human development outcomes for migrants* (Human Development Research Paper Series). New York: United Nations Development Programme.

Sadiq, K. (2005). When states prefer non-citizens over citizens: Conflict over illegal immigration into Malaysia. *International Studies Quarterly, 49*(1), 101–122.

Salt, J., & Clarke, J. (2000). International migration in the UNECE region: Patterns, trends, policies. *International Social Science Journal, 52*(165), 313–328.

Sen, A. (1999). *Development as freedom*. Oxford: Oxford University Press.

Seol, D.-H. (2001). Situation of and measures on undocumented foreign workers in Korea. *Shinhak Sasang 113*, 49–75 (in Korean).

Seol, D.-H., Hwang, P.-K., Goh, H. U., & Yang, H.-W. (2005). Survey on arrest of undocumented foreigners and foreigners detention facilities in Korea. In *2005 human rights situation investigation*. Seoul: National Human Rights Commission (in Korean).

Sin Fronteras. (2011). *Crisis en política migratoria mexicana*. Retrieved July 8, 2011, from http://www.sinfronteras.org.mx/index.php?option=com_content&view=article&id=1121:crisis-en-politica-migratoria-mexicana&catid=31:comunicados&Itemid=67

Skeldon, R. (2006). Recent trends in migration in East and Southeast Asia. *Asian and Pacific Migration Journal, 15*(2), 277–293.

Statistics Korea. (2011). *Total population and population growth*. Retrieved July 4, 2011, from http://www.index.go.kr/

Straubhaar, T. (2002). Towards a general agreement on movements of people (GAMP). *Journal of International Peace and Organization, 77*(1–2), 71–96.

Taran, P., & Geronimi, E. (2003). Globalization, labor and migration: Protection is paramount. In *Perspectives on labor migration*. Geneva: International Labor Organization.

UNDP. (2009). *Human development report 2009. Overcoming barriers: Human mobility and development*. New York: United Nations Development Programme.

UNHCR. (2006). *Addressing mixed migratory movements: A 10-point plan of action*. Geneva: UNHCR.

UNODC. (2004). *United nations convention against transnational organized crime and the protocols thereto*. Vienna: United Nations Office on Drugs and Crime.

UNODC. (2011). *World drug report 2011*. Retrieved June 15, 2011, from http://www.unodc.org/unodc/en/treaties/CTOC/countrylist-traffickingprotocol.html

U.S. Department of Homeland Security. (2011). *Estimates of the unauthorized immigrant population residing in the United States: January 2010*. Retrieved July 6, 2011, from http://www.dhs.gov/files/statistics/immigration.shtm

Vasta, E. (2011). Immigrants and the paper market: Borrowing, renting and buying identities. *Ethnic and Racial Studies, 34*(2), 187–206.

Vera López, R. (2010, November 5). *Human rights crisis in Mexico – Hitting a low point after 200 years of violent independence*. Paper presented at the Rafto Symposium 2010: Trading Human Rights for Security? – The Exploitation of Migrants in Mexico's War on Drugs, Bergen.

Webber, F. (2011). How voluntary are voluntary returns? *Race & Class, 52*(4), 98–107. doi:10.1177/0306396810396606.

Wickramasekara, P. (2011). *Circular migration: A triple win or a dead end*. Geneva: ILO.

Zlotnik, H. (2004). *International migration in Africa: An analysis based on estimates of the migrant stock*. Retrieved August 29, 2006, from http://www.migrationinformation.org/USFocus/display.cfm?ID=252

Chapter 10
Climate Change, Migration, and Development

Susan Martin and Koko Warner

Introduction

As early as 1990, the Intergovernmental Panel on Climate Change (IPCC) warned that significant levels of migration could occur as a result of changing climatic conditions (Brown 2008, p. 11). The concept of environmental migration proved to be a controversial one, largely because of the difficulty in measuring the extent to which environmental factors compel people to move. Since the 1980s, when the term *environmental refugee* was coined, experts within the environmental and migration fields have differed in their characterization of the phenomenon.

Some see the environment as a principal cause of population movements, emphasizing the forced nature of the migration (thus, use the term "refugee"), and often project that hundreds of millions of persons will be affected, frequently without differentiating between those who will move short distances to safer ground versus those who will move thousands of miles to new countries. Others, by contrast, raise questions about the models used to generate estimates of those forced to migrate and emphasize that pull factors in destination locations are often more important than push factors at home in determining whether, where, and in what volume people will migrate.

S. Martin (✉)
Institute for the Study of International Migration, Georgetown University, Washington DC, USA
e-mail: martinsf@georgetown.edu

K. Warner
Environmental Migration, Social Vulnerability and Adaptation, United Nations University, Bonn, Germany
e-mail: warner@ehs.unu.edu

I. Omelaniuk (ed.), *Global Perspectives on Migration and Development: GFMD Puerto Vallarta and Beyond*, Global Migration Issues 1, DOI 10.1007/978-94-007-4110-2_10,

Migration experts and international organizations charged with managing different forms of human mobility and humanitarian needs have been particularly concerned about a potential backlash against migrants and misuse of terms like refugee, which is carefully defined in international law, by those who emphasize the likelihood of mass emergency movements.

Although new research is emerging on the relationship between climate change and migration (Jäger et al. 2009; Kolmannskog 2009a, b; Corlett 2010; White 2011; Piguet et al. 2011), without such basic information as how many people are likely to move, from where to where, and for how long, developing an appropriate policy framework is exceedingly difficult. But even where there is a recognition that some form of migration related to environmental change is likely to occur, addressing these movements is hampered by the paucity of policy or institutional responses deemed appropriate to these forms of migration.

Climate change and migration is relatively new to the agenda of the Global Forum on Migration and Development. The Government GFMD discussions in Athens recommended that policy makers "Give serious consideration to the impact of climate change on migration and to joint efforts to face this challenge" and referred to the need for "mainstreaming and integrating migration into development planning processes, including … National Adaptation Plans of Action concerning climate change (NAPAs)."

There were two brief references to climate change in the report of the Civil Society Days in Athens. Roundtable 1, in discussing the root causes of migration, recommended "Send[ing] a clear message from the GFMD to the Copenhagen Conference on climate change that the threats to and interests of migrants and potential migrants should be high on the policy agenda of the conference" (Global Forum on Migration and Development 2010). Roundtable 3, in discussing data, research, and policy coherence, noted "Climate change must increasingly be taken into consideration in the context of both migration and development" (CSD 2009).

As discussed below, the discussions in Puerto Vallarta were more extensive and demonstrated the complexities of addressing the interconnections between climate change and migration. Following these discussions, in December 2010, the UN Framework for Climate Change Convention (UNFCCC) negotiations took place in Cancun. Termed COP 16, these discussions included workshops on climate change, migration, and displacement and resulted in conclusions and recommendations in the final negotiated document of the conference.

This chapter begins with a discussion of the potential impacts of climate change on migration patterns in both developed and developing regions, citing four possible scenarios. It refers to impacts that are specific to climate change as well as those that result more generally from environmental factors, many of which will be exacerbated as a result of climate change. The chapter then discusses the implications of these scenarios for policy making, with particular attention to development of adaptation, disaster risk reduction strategies, and development policies, on the one hand, and immigration policies in destination countries, on the other. The chapter also discusses the institutional implications resulting from the interconnections among these environmental, development, and migration policies. The following section presents the discussions about climate change, migration, and development at the GFMD in Puerto Vallarta and the follow-up at the COP 16 talks in Cancun. The chapter concludes with recommendations.

How Will Climate Change Affect Migration in Developed and Developing Regions?

There are four paths, in particular, by which environmental change may affect migration either directly or, more likely, in combination with other factors:

1. Changes in weather patterns that contribute to longer-term drying trends that affect access to essential resources such as water and negatively impact the sustainability of a variety of environment-related livelihoods including agriculture, forestry, fishing, etc.
2. Rising sea levels and glacier melt that cause massive and repeated flooding and render coastal and low-lying areas uninhabitable in the longer-term
3. Increased frequency and magnitude of weather-related acute natural hazards, such as hurricanes and cyclones, that destroy infrastructure and livelihoods and require people to relocate for shorter or longer periods
4. Competition over natural resources that may exacerbate pressures that contribute to conflict, which in turn precipitates movements of people (Raleigh et al. 2008; Renaud et al. 2007; Brown 2008; Hugo 2008; Kniveton et al. 2008; Martin 2009)

The first two scenarios are likely to cause slow-onset migration, in which people seek new homes and livelihoods over a lengthy period of time as conditions in their home communities worsen. The third and fourth scenarios are likely to create conditions that cause large-scale displacement, often in emergency situations.

Vulnerability or resilience to these situations—that is, the capability to cope or adapt to them—will determine the degree to which people are forced to migrate. The availability of alternative livelihoods or other coping capacities in the affected area generally determines the scale and form of migration that may take place. For slow-onset events, such as intensified drought and rising sea levels, as compared to such natural disasters as hurricanes, cyclones, and typhoons, or human disasters, such as conflict, the urgency to migrate may be less pressing since the environment and the harm associated with it change more slowly. But if alternative livelihoods are not available within a reasonable timeframe for the would-be migrant, then migration may be the best or only option available, even in slow-onset situations.

Depending on the specific situation, migrants from climate change may resemble labor migrants, seeking better livelihood opportunities in a new location, or they may resemble refugees and internally displaced persons who have fled situations beyond their individual control. Most migration occurring from climate change is likely to be internal, with people seeking better opportunities within their own borders, although a portion will undoubtedly be international. In the most extreme cases, particularly in the context of rising sea levels, an entire population of island nations may need to be resettled. The prospect for such resettlement has been raised by a number of low-lying small island nation states in the Pacific.

Some migration may be temporary while others may be permanent. The decision as to whether return is possible involves a range of variables, including the extent to which the environmental causes—either direct or through other channels—are likely to persist. Policies in the receiving communities and countries, depending on whether the migration is internal or international, will also affect the likelihood for return or settlement in the new location. In addition to immigration policies, the policies affecting return and settlement include land use and property rights, social welfare, housing, employment, and other frameworks that determine whether individuals, households, and communities are able to find decent living conditions and pursue adequate livelihoods. Return and reintegration is also affected by plans and programs to mitigate future dislocations from environmental hazards, coming full circle on the life cycle to a focus on prevention, adaptation, and risk reduction.

Developing countries with a large proportion of people directly involved in agriculture, herding, and fishing are particularly sensitive to environmental changes and to natural disasters. As Collinson (2010) suggests, "many of the world's poorest and most crisis-prone countries will be disproportionately affected by climate change owing to higher exposure to climate-related hazards such as droughts and floods, preexisting human vulnerabilities and weak capacities for risk reduction measures" (p. 5).

However, even highly destructive natural hazards will not necessarily result in humanitarian crises that cause massive displacement. Generally, the efficacy of national and international policies, institutions, and humanitarian responses influences whether people are able to cope with the aftereffects of natural hazards in a manner that allows them to recover their homes and livelihoods. In Bangladesh, for example, natural hazards such as cyclones are only one part of the broader migration picture (Martin and Collinson 2010). As Collinson further observes, "national governance structures and the state's capacity to provide services and maintain institutions are of key importance in influencing the humanitarian impacts of climate change and associated displacement and migration in different contexts" (Collinson 2010, p. 2).

Conflict may further undermine the capacity of people to cope with climate change, leading to greater displacement than might have been the case in a more stable environment. Development is often delayed or even reversed in the context of prolonged conflict, leading to greater vulnerability and decreased capacity to cope with the effects of climate change.

What type of migration occurs as a result of either slow or rapid-onset situations will depend, at least in part, by preexisting institutional experience and capacity to address the needs of affected people as well as preexisting migration patterns. In Jalisco and Zacatecas, Mexico, for example, weather cycles, in combination with other push and pull factors (particularly US labor demand), have contributed to a century of migration to the United States and to urban centers in Mexico (Martin 2010b, c). These Mexican states lie at a crossroads between arid and arable land and have experienced drought cycles that prompted population movements.

Understanding the history of this migration is essential to understanding the ways in which further desertification due to climate change may affect future movements.

Phillip Martin (2010) argues that even without climate change, the coming years are likely to witness continuing large-scale migration out of the agricultural sector, particularly in developing countries where farm incomes are significantly lower than nonfarm incomes. Climate change, specifically global warming, is likely to accelerate this pace of migration. Several economic models project that global warming will have more effects on the distribution of farm production than global farm output, with new areas becoming viable for farming as a result of higher temperatures. However, far more people are likely to be displaced—particularly in developing regions and between developing and developed regions—by global warming than those likely to find jobs in these new farming areas—often located in developed countries where mechanization of new farm operations is probable.

Past and current unsustainable development practices compound problems and pose more immediate challenges than global climate change. In St. Louis, Senegal, for example, a combination of a large dam system on the Senegal River and a poorly planned flood protection canal for St. Louis have contributed to significant coastal erosion that threatens to inundate and displace several small communities (Warner 2010). The canal—originally just 4 m wide—has grown to more than a kilometer wide, with the ocean threatening not only coastal communities but also barrier beaches in a national park. Local fishing and farming communities are losing their livelihoods as saltwater intrudes, contributing to internal migration pressures.

In Jalisco, Mexico, many farmers switched from traditional crops to agave as demand for tequila rose, only to experience falling prices as soil was eroded by new pesticides. Similarly, pollution from urban areas and livestock is affecting both the quantity and quality of water in the Lake Chapala-Lerma-Santiago river basin (Martin 2010b, c). In Bangladesh, the rising level of riverbeds is due in part to changed land and water use upstream (Martin and Collinson 2010).

Climate change increases preexisting vulnerability by allowing more extreme weather-related shocks to affect highly fragile livelihood systems and vulnerable populations. Sea-level rise, for example, worsens already extreme pressures on land due to high population density and growth (Martin and Collinson 2010).

Living in a mid-income country with relatively low fertility rates, Mexicans may be less vulnerable to some kinds of environmental changes, in part because many people have the assets and networks needed for migration. On the other hand, many residents of Bangladesh and Senegal display high levels of vulnerability in terms of poverty, rapid population growth, landlessness, and loss of livelihoods, but those very factors limit the availability of assets that can be used to migrate. Such very high levels of vulnerability may mean that migration, particularly that resulting from acute natural disasters, will occur with little forewarning and under emergency conditions that bode ill for the security and well-being of those who are displaced.

Many international movements from both slow- and rapid-onset situations are likely to be immediately across borders, from one poor, developing country into another. Such migration may be particularly challenging as the receiving countries will likely have few resources, legal structures, or institutional capacity to respond to the needs of the migrants. Geographical proximity may also mean that destination areas face some of the same environmental challenges as areas of migration origin (e.g., drought, desertification) and may offer little respite in this regard.

Implications for Policy Making

The potential scenarios presented above have implications for several areas of policy making: adaptation and disaster risk reduction strategies to address emigration pressures, immigration policies in potential destination countries, and development policies. These are discussed in the following sections.

Adaptation and Disaster Risk Reduction

Adaptation and disaster risk reduction can either (a) involve steps to reduce the need for individuals to migrate to get out of harm's way or (b) involve migration as an adaptation or risk reduction strategy, which allows a community or household to cope with changes and, perhaps, reduce risk for others. *Adaptation* refers to "initiatives and measures to reduce the vulnerability of natural and human systems against actual or expected climate change effects" (IPCC 2007). *Disaster risk reduction* involves "systematic efforts to analyse and manage the causal factors of disasters, including through reduced exposure to hazards, lessened vulnerability of people and property, wise management of land and the environment, and improved preparedness for adverse events" (UNISDR 2009). Identifying vulnerabilities is essential in each case since the "characteristics and circumstances of a community, system or asset … make it susceptible to the damaging effects of a hazard" (UNISDR 2009).

Each of these processes is an integral part of development. Individuals and communities have greater capacity to cope with crises when they have greater human, social, and financial resources to seek and enjoy alternatives.

The Copenhagen Accord adopted at the 15th session of the Conference of the Parties to the UN Framework Convention on Climate Change in December 2009 highlighted the importance of adaptation strategies:

> Adaptation to the adverse effects of climate change and the potential impacts of response measures is a challenge faced by all countries. Enhanced action and international cooperation on adaptation is urgently required to ensure the implementation of the Convention by enabling and supporting the implementation of adaptation actions aimed at reducing vulnerability and building resilience in developing countries, especially in those that are particularly vulnerable, especially least developed countries, small island developing States and Africa. We agree that developed countries shall provide adequate, predictable and sustainable financial resources, technology and capacity-building to support the implementation of adaptation action in developing countries. (Copenhagen Accord 2009)

National adaptation programs of action (NAPAs) are, to date, the principal mechanisms through which low-income developing countries identify adaptation needs and programs. The parties to the accord pledged to provide US $30 billion for the period from 2010 to 2012, with funding allocated between adaptation and mitigation. The most vulnerable developing countries, such as the Least Developed Countries (LDCs), small island developing states, and African countries, will be given priority for adaptation programs. Developed countries also committed to a goal of mobilizing jointly $100 billion dollars a year by 2020 to address the needs of developing countries. To what extent projects related to migration will be funded is unknown.

Adaptation strategies related to migration fall into two major categories. First, and more commonly, governments view adaptation as a way to reduce migration pressures and allow people to remain where they are by modifying agricultural practices, management of pastoral lands, infrastructure such as dykes and coastal barriers, fishing patterns, and other strategies to reduce pressures on fragile ecosystems. Bangladesh, for example, in its NAPA seeks to combat salinization, to slow the "social consequences of mass scale migration to cities." Other approaches focus on early warning and emergency preparedness to reduce displacement from natural disasters associated with climate change.

Second, migration is seen as an adaptation strategy itself. This perspective appears in three contexts. First, some countries see migration as a way to reduce population pressures in places with fragile ecosystems. Second, countries recognize that resettlement of some populations may be inevitable but should be accomplished with proper planning. Third, migrants already living outside of vulnerable areas are seen as potential resources to help communities adapt and respond to climate change. Just as migrants are contributing to the broader development of their home countries, such adaptation strategies envision that the diaspora may have the technical knowledge and financial resources to help communities cope with the effects of climate change.

Migration can be planned or spontaneous, involving individuals and households or entire communities. It can be internal, with people moving shorter or longer distances to find new homes and livelihoods within their own countries, or it can be international, with people seeking to relocate to other countries. It can proceed as an orderly movement of people from one location to another, or it can occur under emergency circumstances. It can be temporary, with most migrants expecting to return home when conditions permit, or it can be permanent, with most migrants unable or unwilling to return. Each of these forms of migration requires significantly different approaches and policy frameworks.

Migration can have positive as well as negative consequences for the affected populations and the communities to which they migrate. The negative impacts stem particularly from emergency mass movements, generally those related to the rapid-onset natural disasters and to competition for resources that may result in conflict. These movements most closely resemble refugee movements and often require large-scale humanitarian assistance. The negative impacts may also be more extreme if receiving communities, particularly urban areas, are unprepared to absorb large numbers of spontaneous migrants. The more positive impacts occur when migration is a voluntary coping strategy that allows people time to weigh alternatives and use migration as a way to reduce household risk (Morton et al. 2008, pp. 5–7).

There are implications as well for migration-related development policies. In recent years, the GFMD and other forums have raised the importance of integrating migration issues more thoroughly into Poverty Reduction Strategy Papers, assessments of progress in achieving Millennium Development Goals, and other development planning processes. The adequacy of initiatives to address migration in development planning has been a principal focus of roundtables since the GFMD in Brussels and is the principal focus of the GFMD Working Group on Policy Coherence, Data and Research. As a direct result of these GFMD discussions, the Global Migration Group produced *Mainstreaming Migration into Development Planning: A handbook for policy-makers and practitioners.* Chapter 3.13 focuses upon environmental issues, explaining that

> environment can be considered a constituent of development in the sense that a) environmental conditions affect quality of life, health and wellbeing of communities, and b) livelihoods, industries and systems of production depend on so-called 'ecosystem services,' ranging from the provision of resources such as fertile soils, mineral and water resources to the capacity of ecosystems to absorb carbon dioxide or withstand erosion. (Global Migration Group 2010)

The handbook urges policy makers and practitioners to "Plan ahead in order to address potential migratory consequences of environmental change, including as a result of climate change," and it makes more specific suggestions about integration of migration concerns into adaptation and disaster risk reduction strategies (Global Migration Group 2010).

International Migration Policies

Immigration laws of most destination countries are not conducive to receiving environmental migrants, unless they enter through already existing admission categories (Leighton 2010; Martin 2010a; McAdam 2011). Typically, destination countries admit persons to fill job openings or to reunify with family members. Employment-based admissions are usually based on the labor market needs of the receiving country, not the situation of the home country. Family admissions are usually restricted to persons with immediate relatives (spouses, children, parents, and, sometimes, siblings) in the destination country.

Humanitarian admissions are generally limited to refugees and asylum seekers, that is, those who fit the definition in the UN Convention Relating to the Status of Refugees: persons with a well-founded fear of persecution on the basis of race, religion, nationality, membership in a particular social group, or political opinion. Most environmental migrants will be unlikely to meet the legal definition of a refugee under national or international law because they would not be able to demonstrate that the cause of their displacement is related to persecution.

Some countries have established special policies that permit individuals whose countries have experienced natural disasters or other severe upheavals to remain at least temporarily without fear of deportation. The United States, for example, enacted legislation in 1990 to provide temporary protected status (TPS) to persons "in the United States who are temporarily unable to safely return to their home country because of ongoing armed conflict, an environmental disaster, or other extraordinary and temporary conditions." Environmental disaster may include "an earthquake, flood, drought, epidemic, or other environmental disaster in the state resulting in a substantial, but temporary, disruption of living conditions in the area affected."[1] In the case of environmental disasters, as compared to conflict, the country of origin must request designation of TPS for its nationals.

Importantly, TPS only applies to persons already in the United States at the time of the designation. It is not meant to be a mechanism to respond to an unfolding crisis in which people seek admission from outside the country. It also only pertains to situations that are temporary in nature. If the environmental disaster has permanent consequences, a designation of TPS is not available, even for those presently in the United States, or it may be lifted. When the volcano erupted in Montserrat in 1997, TPS was granted to its citizens and was extended six times. In 2005, however, it was ended because "it is likely that the eruptions will continue for decades, [and] the situation that led to Montserrat's designation can no longer be considered 'temporary' as required by Congress when it enacted the TPS statute" (U.S. Federal Register 2004). Given the temporary nature of the grant and its application only to those already in the country, TPS has limited utility in addressing climate change-related migration.

At the European Union level,

> the Temporary Protection Directive establishes temporary protection during "mass influxes" of certain displaced persons. The term "mass-influx" refers to situations where masses of people are suddenly displaced and where it is not feasible to treat applicants on an individual basis. It was decided that "mass-influx" was to be defined on a case-by-case basis by a qualified majority of the Council. (Kolmannskog 2009b, p. 4)

Sweden and Finland have included environmental migrants within their immigration policies. Sweden includes within its asylum system persons who do not qualify for refugee status but have a need for protection. Such a person in need of protection "has left his/her native country and does not wish to return there because he or she:—has a fear of the death penalty or torture—is in need of protection as a result of war or other serious conflicts in the country—is unable to return to the native country because of an environmental disaster."[2] The decision is made on an individual, not group basis. Although many recipients of this status are presumed to be in temporary need of protection, the Swedish rules foresee that some persons

---

[1] Public Law 101–649 (Immigration Act of November 29, 1990), Section 302.

[2] Aliens Act (2005: 716), Issued September 29, 2005, with amendments: up to and including Swedish Code of Statutes 2009:16; Chapter 4. Refugees and persons otherwise in need of protection.

may be in need of permanent solutions. Similarly, in the Finnish Aliens Act, "aliens residing in the country are issued with a residence permit on the basis of a need for protection if … they cannot return because of an armed conflict or environmental disaster" (Kolmannskog 2009b, p. 4).

A number of countries provide exceptions to removal on an ad hoc basis for persons whose countries of origin have experienced significant disruption because of natural disasters. After the 2004 tsunami, for example, Switzerland, the UK, and Canada suspended deportations of those from such countries as Sri Lanka, India, Somalia, Maldives, Seychelles, Indonesia, and Thailand. A number of governments announced similar plans after the 2010 earthquake in Haiti. However, there is no international law that would compel or even encourage other governments to follow similar policies.

To date, there are no examples of legislation or policies that address migration of persons from slow-onset climate changes that may destroy habitats or livelihoods in the future.[3] For the most part, movements from slow-onset climate change and other environmental hazards that limit economic opportunities are treated in the same manner as other economically motivated migration. Persons moving outside existing labor and family migration categories are considered to be irregular migrants. In the absence of a strong humanitarian basis for exempting them from removal proceedings (which is unlikely in the slow-onset situation), these migrants would be subject to the regular systems in place for mandatory return to their home countries. Because their immediate reasons for migrating would be similar to that of other irregular migrants—that is, lack of economic opportunities at home and better economic opportunities abroad—there would be little reason for destination countries to manage these movements outside their existing immigration rules.

Institutional Arrangements

Just as the legal frameworks for addressing climate change-induced cross-border migration are weak, so are the institutional roles and responsibilities at both the international and the national levels. With the exception of the refugee regime, in which clear responsibility is given to the UNHCR, there is no existing international regime for managing international movements of people, let alone those compelled to move by the effects of climate change.

[3] New Zealand's Pacific Access Category (PAC) is sometimes described as climate change and migration legislation. Under PAC, 75 people from Tuvalu, 75 from Kiribati, and 250 from Tonga (islands subject to rising sea levels) may immigrate to New Zealand each year. The program is based on employment, however, not environmental factors. The immigrants must be between 18 and 45 years old, have an offer of employment in New Zealand, have English skills, meet minimum income requirement, undergo a health check, and have no history of illegal entrance. The program is not intended to provide access to those who may be most vulnerable to climate change-induced displacement—the elderly or the infirm, for example.

This is not to say that there is a total absence of governance. There is a plethora of international organizations that have some responsibilities for international migration, also in the context of environmental change. Perhaps the most important international organization in this area is the International Organization for Migration (IOM). IOM has been a focal point for discussion of environmental migration since 1992, when it cohosted a series of consultations on the interconnections between the environment and migration, in the context of the UN Conference on Environment and Development (UNCED) in Rio de Janeiro. Since then, IOM has published working papers and books on climate change and migration and worked with UN agencies and experts to define more precisely the category of environmental migration.[4]

The UN High Commissioner for Refugees has also raised questions regarding international responsibility for climate change-induced migration, but its own role would likely focus primarily on emergency movements stemming from conflict-related displacements. The International Labor Organization has focused on climate change mitigation and adaptation in the context of its commitment to decent work standards, observing "The world can ill afford to invest the massive resources required to address the climate crisis in ways that do nothing to tackle the global job crisis and poverty" (ILO 2007).

The Global Migration Group (GMG), established to promote coordination and identify gaps in the international system, has also focused some attention on climate change and migration. The GMG has listed a number of relevant documents pertaining to these issues, including reports from a number of the member organizations. Yet absent from the GMG are the UN Environment Program (UNEP)[5] and the Secretariat of the UNFCCC, which have significant interest in these issues. This is unfortunate because UNEP has collaborated with members of the GMG on specific projects. Their involvement in discussions on migration is at least as relevant as those of the other specialized agencies that participate in the GMG. To address concerns that the GMG has grown too large, and has too many members with peripheral interest in migration, it might make sense to invite specialized agencies only when issues connected to their core agendas are to be discussed in the GMG.

---

[4] For example, see Laczko and Aghazarm (2009).

[5] UNEP's program related to climate change and migration has three principal objectives:

1. Research, hot spot identification, and assessment of high priority vulnerable countries and subnational regions for targeted UNEP assistance on ecosystems management and climate change adaptation

2. Awareness raising in the international humanitarian and development community of the importance of environmental sustainability in rural development and its role in forced migration movements

3. Country project development to assist in policy direction, plans, and programs in identified priority countries and subnational regions toward ecosystems management and climate change adaptation

GFMD Engagement with Environment, Migration, and Development

When the government of Mexico assumed leadership of the GFMD discussions for 2010, they placed the issue of climate change, migration, and development more firmly on the agenda in a roundtable session of its own. Responsibility for the government roundtable was assumed by the United Kingdom and Bangladesh, with support from Chile, Ecuador, France, Germany, Ghana, Mauritius, Mexico, Switzerland, and IOM, and academic experts were requested to prepare the background paper.[6]

The GFMD meetings generally result in rapporteur reports and not negotiated text of an agreement. Nevertheless, the rapporteur generally seeks to report on areas in which there was a great deal of consensus. The rapporteur's report on the climate change, migration, and development discussions in Puerto Vallarta touched on the following four major points:

1. Data and analysis
   The discussions acknowledged conceptual difficulties and lack of empirical data showing a definitive link between climate change and migration but recognized that "this is an area deemed too important to ignore." The roundtable emphasized the need to collect data and bring them together on a continuous and systematic basis.
2. Geographical scale
   The UK paper had made the now common assertion that most displacement from climate change is likely to be internal (i.e., within the borders of the affected countries) indicating that discussions of climate change and migration might be of relatively less importance for a forum that focuses primarily on international migration and development. But this does not necessarily apply to small island states, which immediately countered this assertion by pointing out that there is no interior in many of their countries to which affected populations could resort in the face of rising sea levels.

   The rapporteur report acknowledged that geography and size do matter in developing responses to climate change displacement. "Small island countries are different: for them internal migration is not an option." Thus, a need to develop this topic at local, regional, and global level was acknowledged. As one representative very eloquently put it, "climate change does not stop at borders."
3. Migration and climate change as a multisectoral issue
   The discussions recognized that these issues cross-cut different policy areas, "not the least of which food and water security, the basic factors of life." The roundtable emphasized the need to discuss migration and displacement from climate change in multiple forums, including the UNFCCC, and to integrate these issues into National Adaptation Programs of Action (NAPAs), Disaster Risk Reduction Strategies (DRRs), and Poverty Reduction Strategy Papers.

[6] The background paper was written by Ronald Skeldon of the UK Department for International Development (DFID) with input from Susan Martin at Georgetown University. Martin and Warner wrote the background paper for the parallel discussions that took place during the Civil Society Days.

4. Legal and institutional arrangements
   The roundtable consensus was that new instruments and institutional arrangements were needed to tackle the complex issues arising from climate change-induced migration and displacement, but it was recognized that binding agreements are the result of complex negotiations. The rapporteur's report indicated: "One must deal with complex issues, such as development, migration, humanitarian issues, and climate change which are interconnected, while at the same time trying to achieve as well policy coherence in all of them."

In keeping with the nonbinding nature of the discussions, the roundtable recommended the following ways to follow up on the discussions:

1. In order to expedite data and analysis exchange and sharing, first experiences and best practices are of the utmost importance. A virtual library may be a useful way of sharing this information.
2. A need exists to strengthen the dialogue at local, regional, and global level on the interconnections of climate change and migration. RT participants welcomed and encouraged future discussions on this issue in the context of the Global Forum.
3. The need to begin discussions on the appropriate legal and institutional arrangements to address these important issues was recognized. (GFMD 2010, p. 3)[7]

Although the conclusions of the GFMD are nonbinding, there has been follow-up action on some of these recommendations from 2010, particularly in regard to discussions on appropriate legal and institutional arrangements. For example, in June 2011, the Nansen Conference on Climate Change and Displacement in the twenty-first century was held in Oslo. About 200 participants, representing governments, nongovernmental organizations, civil society, research institutions, intergovernmental organizations, and the private sector, attended the conference. Chaired by Margareta Wahlström, UN Special Representative of the Secretary-General for Disaster Risk Reduction, the meeting produced a set of draft principles on climate change and displacement for consideration at the UN High Commissioner on Refugees' Ministerial Conference to be held in December 2011.

Climate Change and Migration at COP 16

From 2008 onward, migration and displacement have been on the agenda of the climate change negotiations undertaken in the context of the UN Framework Convention on Climate Change, known by their acronym COP. The discussions have been largely within the context of negotiation of text regarding adaptation policies and programs. After the perceived failure of the COP15 negotiations in Copenhagen, which called into question the efficacy of climate negotiations in a UN forum, there

[7] GFMD (2010). "Roundtable 3: Policy and institutional coherence to address the relationship between migration and development." General Rapporteur's Report to Plenary. Available at http://www.gfmd.org/en/documents-library/mexico-2010.html

was pressure to create a package of balanced outcomes for Cancun, which would not be too ambitious but would deliver concrete enough results to restore faith in the UNFCCC process. In this context, delegates focused increasingly on what kinds of elements could be included in a possible Cancun Adaptation Framework.

The language on migration and displacement was placed in the context of a broader range of adaptation issues found in Article 14. Article 14 begins with a general statement and specifies actions related to migration in subsection (f):

> Invites all Parties to enhance action on adaptation under the Cancun Adaptation Framework, taking into account their common but differentiated responsibilities and respective capabilities, and specific national and regional development priorities, objectives and circumstances, by undertaking, inter alia, the following: …
>
> (f) Measures to enhance understanding, coordination and cooperation with regard to climate change induced displacement, migration and planned relocation, where appropriate, at national, regional and international levels.

The negotiating language focused on enhancing understanding of the phenomenon and increasing coordination and cooperation, rather than offering concrete solutions. The negotiations reinforced that the issue was important but not controversial in terms of what was being asked: voluntary measures to enhance understanding, coordination, and cooperation. While the topic of migration and displacement itself has the potential to be divisive, the way that it had been couched and presented to UNFCCC delegates (voluntary, not embedded in normative language, not linked to contentious issues) prepared the grounds for its inclusion.

At COP 16, countries committed to providing funding for adaptation, so an advantage of Article 14(f) is that those items mentioned in the Cancun Adaptation Framework are viewed in practical terms by many ministries as types of activities that may qualify for adaptation funding, which at present is emerging in a variety of funds and sources, as well as traditional bilateral assistance. Yet another advantage of discussing displacement in the UNFCCC context is that parties see the linkages between mitigation, development, humanitarian issues, and (at least for some) potential (human) security issues in the future. These links encourage cross-ministry policy approaches to tackle complex issues like migration and displacement. It is too soon to know, however, whether governments will follow-up with concrete proposals and a firm commitment to funding projects that focus on displacement, migration, and planned relocation.

Moving Ahead

Discussion of laws, policies, and organizational responsibilities to manage environmental migration is in its infancy. As understanding increases of the various ways that environmental change affects migration patterns, and vice versa, governments, civil society representatives, and experts are beginning to think through how to manage the implications of these interconnections. Much of the attention to date focuses legitimately (given that most movements are likely to be within countries) on internal migration, largely in the context of adaptation strategies and, to a lesser degree, movements that may arise as a result of natural disasters and climate change-induced conflict.

Nevertheless, some international migration may well be needed, particularly for the citizens of island nations, necessitating identification of appropriate admissions policies in potential destination countries. Highest priority should be given to identifying likely patterns of migration. Particular attention should be placed on identifying who cannot be relocated within their home countries, either because of widespread habitat destruction (again, as in the case of certain island states) or because relocation would pose security risks that could provoke violence or even conflict. Some attention should also be addressed to the slow-onset situations in which loss of livelihoods generates emigration pressures. In the absence of legal opportunities to immigrate, at least some portion of those who lose livelihoods as a result of climate change and other environmental hazards may become irregular migrants.

Few potential destination countries have explicit policies to manage climate change-induced migration, unless affected populations migrate through the normal immigration policies that give preference to family reunification and employment-based admissions. While potential destination countries have asylum or resettlement systems to manage admission of persons who cannot return home because of a well-founded fear of persecution, none have systems in place to manage admission of persons who cannot remain or return home because of environmental threats. At best, destinations of international migrants have policies to defer deportation of those coming from countries with natural disasters, but these are generally post disaster and ad hoc in their implementation.

In sum, no major destination country has a proactive policy designed to resettle persons adversely affected by environmental hazards.

Given the current gaps, more attention needs to be placed on identifying and testing new frameworks for managing potential movements. Although climate change and migration is now on international agendas, there are still few examples of concrete projects or policies to address the interconnections between the two phenomena. Moreover, the links to development are not yet well understood. Attention needs to be given to both sides of the environment and migration nexus: identifying adaptation strategies that allow people to remain where they currently live and work and identifying migration and resettlement strategies that protect people's lives and livelihoods when they are unable to remain. Since internal migration is the most likely outcome for those affected by climate change and other environmental hazards, highest priority should be given to policies and programs aimed at managing these issues within the most affected countries, particularly in the least developed states that are particularly vulnerable to the impacts of climate change.

A principal challenge for potential destination countries is determining whether those who migrate because of environmental stress should be given consideration over others who migrate in search of better opportunities. There is reason for skepticism that many destination countries will answer this question in the affirmative. With the exception of their refugee and asylum policies, countries tend to frame their admissions policies around their own national interests, prioritizing admission of persons who will contribute to economic growth, meet labor shortages, or have close family ties in the destination country. While exceptions may be made for environmentally induced migrants whose situation most resembles that of refugees, there is less likelihood that governments will make an exception for those who resemble economic migrants. Yet at least in the case of island states that are affected by rising sea levels, the absence of an appropriate international response could render large numbers of people effectively stateless because they are unable to remain at home but have no legal recourse for entry elsewhere.[8]

In moving toward more coherent frameworks, the lessons of the past will be useful, particularly for those countries that foresee the possibility that planned resettlement, including international movements, may be needed. More systematic examination of involuntary resettlement programs undertaken in the context of development projects would help ensure that climate change-induced resettlement programs do not fall victim to problems identified in these initiatives.

Michael Cernea cites eight interrelated risk factors associated with resettlement from development projects: landlessness, joblessness, homelessness, marginalization, food insecurity, increased morbidity and mortality, loss of access to common property, and social disintegration (Cernea 2000). Under the worst case scenarios, when the long-term needs of the relocated have not been taken into account, the displaced have been at serious risk of "becoming poorer than before displacement, more vulnerable economically, and disintegrated socially" (Robinson 2003).

Identification of best case examples of resettlement—that is, programs that respect the rights of the resettled and resulted in an improved economic and social situation—is as important as identification of pitfalls experienced in programs that failed. World Bank and other guidelines promulgated to protect those who are involuntarily resettled from development projects or who are displaced from natural disasters should be examined systematically to determine their applicability to the resettlement programs proposed by countries fearing the worst consequences of climate change. Lessons from successful refugee resettlement programs may also be useful in thinking through options for resettlement in the context of climate change.

Technical assistance and training for the ministries that may be responsible for resettlement is essential to ensure that all alternatives are exhausted before people

[8] The inadequacies of temporary protection, as evidenced in the Montserrat case study, highlight the potential problems that these situations present. It should be noted that UNHCR does have a mandate for statelessness and has reported some success in reducing its incidence in the past few years. In discussing statelessness, however, UNHCR does not refer to climate change as a potential cause; rather, it cites the following reasons: "Statelessness occurs for a variety of reasons including discrimination against minority groups in nationality legislation, failure to include all residents in the body of citizens when a state becomes independent (state succession) and conflicts of laws between states." See UNHCR, "Searching for citizenship", Available at http://www.unhcr.org/pages/49c3646c155.html. Also, see UNHCR (2009).

are required to relocate, affected populations are involved in the planning, and all steps are taken to ensure appropriate preparations and implementation.

The promulgation of such guidelines and the development of policies to respond to environmental migration will require new modes of international cooperation. As noted above, the need for enhanced international, regional, and local coordination on the issue is noted in the UNFCCC Adaptation Framework adopted in Cancun.

Given the weak legal and institutional situation today, and the looming prospect that climate change will significantly contribute to increased internal and international migration, efforts to build an effective system of governance should begin soon. Whether a new convention on climate change-related displacement is desirable or, for that matter, feasible are questions that beg easy answers.[9] The history of international conventions related to migration is a mixed one, determined largely by whether the movements are forced or voluntary. While the refugee convention and trafficking protocol are widely ratified, the conventions on labor migrants have had low levels of ratification. Because climate change-induced migration will likely have elements of both forms of migration, depending on whether the trigger is slow or rapid onset, the future of such a convention would be questionable.

Beyond feasibility, a number of other issues would need to be addressed before determining that a new convention is the best way to improve governance related to climate change-induced migration. First, how should climate change migrants be defined? For that matter, what term should be used in categorizing this form of migration, particularly given the other factors that are usually at play in explaining movements? Would the same approach work for internal and international movements? Who among those who migrate because of climate change or related environmental reasons are deserving of international protection? And, the list goes on.

The promulgation of the Guiding Principles on Internal Displacement may be a more effective model for developing international standards related to climate change-induced migration across international borders. The Guiding Principles built on existing human rights, humanitarian, and refugee law to develop norms and standards that apply to internally displaced persons. Although they are not themselves binding, the Guiding Principles bring together pertinent law in a manner that has allowed governments and international organizations to set standards that are consistent with international norms. The adoption of the AU Convention on Internally Displaced Persons, which is based largely on the Guiding Principles, demonstrates that this approach has given credibility to the norms established, although the convention has not yet entered into force. It is particularly relevant for this chapter that the AU convention has incorporated protection of those displaced by climate change into its provisions. A similar process that would also address international movements could be equally effective in developing a legal framework for addressing climate change and migration.

[9] For arguments in favor of a convention, see Hodgkinson et al. (2010).

Concrete Steps to Move Ahead

Governments and civil society need to make strategic investments in four areas that would help realize the solutions mentioned above:

1. *Science.* Governments and the research community should support more in-depth qualitative and quantitative research in specific hot spot regions; the collection of necessary demographic, socioeconomic, and environmental data (such as through censuses); and rigorous research to understand the different patterns and scenarios of climate change, migration, and displacement in specific areas.
2. *Dialogue and exchange of best practices.* Governments should foster solutions-oriented policy dialogues that review existing experience and identify emerging good practices in areas such as designing alternative livelihoods for those displaced or threatened by climate change, facilitating migration where appropriate, relocation, and resettlement of populations. There is little time to waste—earnest policy dialogue should begin now when there is still space and time to navigate some of the challenges and opportunities that arise with environmentally induced migration.
3. *Participatory policy planning.* As governments plan responses to climate change and migration, they should involve affected communities in policy planning and implementation of human mobility solutions. In some cases, this may mean site identification for relocation projects; in other instances, it may mean development of alternative livelihoods or agricultural practices which might ease pressure on migration. Governments should also recognize that migrants and diasporas can be effective partners in addressing climate change-induced migration and involve them in planning processes. Such participatory processes have been effective in lessening the negative impacts of development programs that have led to involuntary resettlement.
4. *Proactive approaches.* Governments and civil society need to get ahead of the curve. They should support effective adaptation strategies that take potential migration impacts of climate change into account. They should also seek implementation of effective disaster risk reduction, conflict mediation, and disaster management programs to reduce the likelihood of emergency movements with accompanying humanitarian consequences.

References

Brown, O. (2008). *Migration and climate change* (Migration Research Series, Vol. 31). Geneva: International Organization for Migration.

Cernea, M. (2000). Risks, safeguards and reconstruction: a model for population displacement and resettlement. In M. Cernea & C. McDowell (Eds.), *Risks and reconstruction: Experiences of resettlers and refugees*. Washington, DC: World Bank.

Collinson, S. (2010). *Developing adequate humanitarian responses* [Background Papers of the Study Team on Climate Change and Migration]. Washington, DC: German Marshall Fund of the United States.

Copenhagen Accord. (2009). *Following the United Nations Climate Change Conference*. Retrieved December 2009 from http://unfccc.int/resource/docs/2009/cop15/eng/l07.pdf

Corlett, D. (2010). *Stormy weather: The challenges of climate change and displacement*. Sydney: University of New South Wales Press.

Federal Register. (2004, July 6, Tuesday). Termination of the designation of Montserrat under the Temporary Protected Status Program; extension of employment authorization documentation. *Federal Register, 69*(128).

Global Forum on Migration and Development. (2010, November 8–11). *Partnerships for migration and human development: Shared prosperity–shared responsibility: Report of the Proceedings*. Fourth Meeting of the Global Forum on Migration and Development (GFMD), Puerto Vallarta, Mexico. Retrieved from http://gfmd.org/en/documents-library/mexico-2010.html

Global Migration Group. (2010). *Mainstreaming migration into development planning: A handbook for policy-makers and practitioners*. Geneva: GMG.

Hodgkinson, D., Burton, T., Anderson, H., & Young, L. (2010). *'The hour when the ship comes in': A convention for persons displaced by climate change*. Retrieved from http://www.ccdp-convention.com/documents/CCDP_Convention_Summary.pdf

Hugo, G. (2008). *Migration, development and environment* (Migration Research Series, Vol. 35). Geneva: International Organization for Migration.

Intergovernmental Panel on Climate Change (IPCC). (2007). *Impacts, adaptation and vulnerability* (IPCC Fourth Assessment Report: Working Group II Report). Geneva: IPCC, Appendix 1, "Glossary" p. 869. Retrieved from http://www.ipcc.ch/pdf/assessment-report/ar4/wg2/ar4-wg2-app.pdf

International Labour Organization (ILO). (2007). *Green jobs: Facing up to 'an inconvenient truth'* (World of Work No. 60). Retrieved September 26, 2011 from http://www.ilo.org/wow/Articles/lang--en/WCMS_083900/index.htm

Jäger, J., Frühmann, J., Günberger, S., & Vag, A. (2009, May 14). *Environmental change and forced migration scenarios* (Project synthesis report, 2009, deliverable 044468). Retrieved September 26, 2011 from http://www.each-for.eu/documents/EACH-FOR_Synthesis_Report_090515.pdf

Kniveton, D., Schmidt-Verkerk, K., Smith, C., & Black, R. (2008). *Climate change and migration: Improving methodologies to estimate flows* (Migration Research Series, Vol. 33). Geneva: International Organization for Migration.

Kolmannskog, V. (2009a). *Climate change, disaster, displacement, and migration: Initial evidence from Africa* (UNHCR Research Paper, Vol. 180). Geneva: UNHCR.

Kolmannskog, V. (2009b, January). *Climate change-related displacement and the European response*. Paper presented at the Society for International Development (SID), Vijverberg Session on Climate Change and Migration, The Hague.

Laczko, F., & Aghazarm, C. (Eds.). (2009). *Migration, environment and climate change: Assessing the evidence*. Geneva: International Organization for Migration.

Leighton, M. (2010). *Climate change and migration: Key issues for legal protection of migrants and displaced persons*. Washington, DC: German Marshall Fund of the United States.

Martin, S. (2009). Managing environmentally induced migration. In F. Laczko & C. Aghazarm (Eds.), *Migration, environment and climate change: Assessing the evidence*. Geneva: International Organization for Migration.

Martin, P. (2010). *Climate change, agricultural development and migration* (Background Papers of the Study Team on Climate Change and Migration). Washington, DC: German Marshall Fund of the United States.

Martin, S. (2010a). *Climate change and international migration*. Washington, DC: German Marshall Fund of the United States.

Martin, S. (2010b, February). *Jalisco and Zacatecas study tour report*. Unpublished report available from the Institute for the Study of International Migration, Georgetown University.

Martin, P. & Collinson, S. (2010, March). *Bangladesh study tour report*. Unpublished report available from the Institute for the Study of International Migration, Georgetown University.

McAdam, J. (2011, March 29–30). *Climate change, displacement and the role of international law and policy*. Paper prepared for the International Dialogue on Migration 2011. The Future of Migration: Building Capacities for Change Intersessional Workshop on Climate Change, Environmental Degradation and Migration. Retrieved September 26, 2011 from http://www.iom.int/jahia/webdav/shared/shared/mainsite/microsites/IDM/workshops/climate-change-2011/SessionIII-Paper-McAdam-Session.pdf

Morton, A., Boncour, P., & Laczko, F. (2008). Human security policy challenges: Climate change and displacement. *Forced Migration Review, 31*, 5–7.

Piguet, É., Pécoud, A., & de Guchteneire, P. (2011). *Migration and climate change*. Cambridge: Cambridge University Press.

Raleigh, C., Jordan, L., & Salehyan, I. (2008). *Assessing the impact of climate change on migration and conflict*. Washington, DC: World Bank.

Renaud, F., Bogardi, J., Dun, O., & Warner, K. (2007). *Control, adapt or flee how to face environmental migration?* Bonn: United Nations University.

Robinson, C. (2003). *Risks and rights: Causes, consequences, and challenges of development-induced displacement*. Washington, DC: Brookings Institution.

UN International Strategy for Disaster Reduction (UNISDR). (2009). *Terminology on disaster risk reduction*. Retrieved from www.unisdr.org/eng/terminology/terminology-2009-eng.html

UNHCR. (2009). *Statelessness: An analytical framework for prevention, reduction and protection*. Geneva: UNHCR.

Warner, K. (2010, February). *Senegal study tour report*. Unpublished report available from the Institute for the Study of International Migration, Georgetown University.

White, G. (2011). *Climate change and migration*. Oxford: Oxford University Press.

Chapter 11
Assessing the Impact of Migration Policies on Economic and Social Development

Khalid Koser*

Introduction

Assessing the impact of policy is an essential component of good governance, including for migration. Assessment is required to estimate the impact of policy on individual migrants, on the host population, and on development in countries of origin (Commission on International Migration Data for Development Research and Policy 2009). Policy should be examined from a macroeconomic perspective to assess the extent to which it yields net social gains. From a cost-benefit perspective, it is also necessary to assess whether the best possible outcome has been achieved relative to the cost of each policy (Ardittis and Laczko 2008). It is also good practice to learn lessons from previous policy experiences in devising new policy approaches to prepare for future challenges. A recent review of policy responses to financial and economic crises during the twentieth century, for example, has demonstrated that many of the impacts of the more recent global economic crisis on migration replicate earlier impacts, and in some cases, current policy responses have not learned from the mistakes of previous responses (Koser 2009).

More generally, demonstrating a positive impact is important to maintain public confidence in government and in the ability of government to manage migration in particular, especially as migration continues to rise on the political and media agenda.

Yet in most countries, there is insufficient assessment of the effectiveness and impact of policies and programs on migration and integration, and mechanisms that facilitate the exchange of information and lessons learned both within and between states are

*This chapter is based on a paper prepared for Roundtable Session 3.1 of the 2010 meeting of the Global Forum on Migration and Development, cochaired by the Governments of Argentina, Kenya, and Switzerland (Koser 2010a).

K. Koser (✉)
Academic Dean, Geneva Centre for Security Policy, Geneva, Switzerland
e-mail: k.koser@gcsp.ch

I. Omelaniuk (ed.), *Global Perspectives on Migration and Development: GFMD Puerto Vallarta and Beyond*, Global Migration Issues 1, DOI 10.1007/978-94-007-4110-2_11,

lacking. A particular gap is the assessment of the impact of migration policies on development. While development cooperation is one of the most intensively assessed government sectors, the impact (and costs) of migration policies on development is often unknown. This despite the significant recent policy attention paid to the migration-development nexus, for example, through the meetings and working groups of the Global Forum on Migration and Development (GFMD), and the enormous investment of resources in migration projects, programs, and policies targeted on development.[1]

Definitions and Concepts

Part of the challenge of impact assessment in the migration policy context is a lack of consensus on terminology. First, there is a distinction to be made between migration projects, programs, and policies, and impact assessment applies to all three. A project is normally unique and is of definite duration. In the migration policy context, projects are often undertaken by national or international agencies and organizations, for example, to deliver a specific outcome for a specific group of migrants. An example might be a project to provide business setup training for a group of returnees from a specific country in order to assist them reintegrate. A program is normally defined as a portfolio of projects, intended together to achieve an overall goal. In the migration policy context, for example, a program on migrant integration might combine a set of projects concerning specific outcomes (education, housing, health) for a selected group of migrants (children, asylum seekers, disabled). A policy, in contrast, is normally understood as a principle or set of guidelines to guide decisions.

Thus, a national migration policy may set targets as regards the admission of migrants or conditions for migrants to remain and work in the country or to promote circular migration. Rarely does a national migration policy set development-related targets, either in the country of origin or destination.

A second distinction needs to be made between the concepts of monitoring, evaluation, and impact assessment. Monitoring is the systematic collection and analysis of information as a project or program progresses. It is aimed at improving the efficiency and effectiveness of a project's or program's implementation and matches its progress toward achieving the results planned during the planning phases. It helps to keep the work on track and can let managers know when things are going wrong. Monitoring is a "snapshot" in time. Evaluation is the comparison of actual project and program results against the original plans. It considers initial objectives and intended results, the extent to which they have been accomplished, and how

[1] The chapter presents some of the findings of the GFMD ad hoc Working Group on Policy Coherence, Data, and Research which undertook an "Informal inquiry on policy impact assessments," and organized a seminar on *Assessing the impacts of migration and development policies* in Vienna in June 2010, as well as some of the conclusions of the GFMD/IOM/World Bank workshop on *Managing migration for Policymaking, assessment, and evaluation* that took place in Marseille in June 2011.

they were accomplished. It entails a far more in-depth study than monitoring. An earlier study on how the costs and impacts of migration policies are evaluated offered a working definition: "a relatively regulated and systemic exercise looking at the relevance, efficiency, effectiveness, internal and external coherence, synergies, impact, added value, and sustainability of policies or programs" (Arditis and Laczko 2008, p. 2).

Together, monitoring and evaluation allow for impact assessment. Impact assessment assesses the changes that can be attributed to a particular intervention, whether a project, program, or policy. For example, in the health field, Health Impact Assessment (HIA) is defined as "a combination of procedures, methods, and tools by which a policy, program or project may be judged as to its potential effects on the health of a population, and the distribution of those effects within the population" (European Centre for Health Policy ECHP 1999, p. 4). The impact can be in the short or long term, positive or negative, direct or indirect, and intended or unintended. Impact assessment helps answer key questions for evidence-based policy making: what works, what does not work, where, why, for whom, and for how much?

Overall impact assessment is an essential element in the policy cycle, which can be envisaged as progressing through the following stages: agenda setting, policy formulation, adoption, implementation, and impact assessment. When conducted properly, impact assessment can help inform changes to policies to make them more effective and sustainable.

An Overview of Current Practices in Impact Assessment

In total, nine countries responded to the "Informal inquiry on policy impact assessments" conducted by the GFMD ad hoc Working Group on Policy Coherence, Data, and Research in June 2010: Côte d'Ivoire, six members of the European Union (Belgium, France, Denmark, Finland, the Netherlands, and Spain), Moldova, and Switzerland. Plainly, this is not a representative sample of states, and to an extent, the sample may be self-selecting in that perhaps only those states active in the field of impact assessment responded to the inquiry. At the same time, they represent a good range of origin, transit, and destination countries, combining those in more developed and less developed parts of the world. An obvious recommendation from this chapter is that a more extensive survey is required on the topic of impact assessment.

Responses even from this small sample of countries demonstrated a wide range of current practices on assessing the impact of migration policy on development.

First, among the surveyed countries, very few impact assessments of policies that specifically target migration and development are currently taking place, even if more general procedures for assessing the impact of migration policy do exist and may capture developmental impacts occasionally.

Six of the nine countries that responded to the enquiry have established formal policies for impact assessment in the field of migration policy, and three have reported that they are considering developing them. Côte d'Ivoire, for example, has established a task force to develop a national migration policy including a mechanism for monitoring, evaluation, and impact assessment. In Moldova, the Netherlands, and Switzerland, impact assessment is specific to migration and development policies. Moldova has allocated institutional responsibility for monitoring the implementation of its "Action plan on stimulating the return of Moldovan migrants working abroad," "Programme of action for supporting persons born in the republic of Moldova residing abroad," and "National action plan on protecting Moldovan citizens residing abroad."

The Netherlands planned an assessment of the impact of its "International migration and development 2008" Policy Memorandum for the end of 2011. The Swiss Global Program on Migration and Development has developed evaluation criteria for all its projects and programs and plans to assess their impact in a midterm and final assessment. In the other three surveyed countries, impact assessment is targeted on immigration and integration (Finland and Spain) or is part of a wider assessment of development assistance (Denmark).

Second, procedures for impact assessment differ for projects and programs implemented by government and for those funded by government but implemented by external partners. In six of the nine countries (Belgium, Finland, France, Moldova, Spain, and Switzerland), government staff undertakes monitoring, evaluation, and assessment of projects and programs. Of the three countries where government staff was not reported to be involved, there was no response on this question from Côte D'Ivoire; Denmark tends to commission external evaluations, and the Netherlands relies on half-yearly reports and regular discussions with implementing partners.

Third, where impact assessment takes place within government, institutional responsibility and capacity vary significantly. In some of those countries that involve government staff in the process, assessments are undertaken by the specific department or ministry with responsibility for the particular project or program. In Moldova, in contrast, an interdepartmental committee has been established to monitor the implementation of the Joint Declaration on Moldova-EU Mobility Partnership, including representatives from the Ministry of Labour, Ministry of Social Protection, and Ministry of the Family, as well as the Ministry of Foreign Affairs and European Integration.

Similarly, in Switzerland, an interdepartmental Steering Group has been established to monitor assisted voluntary return programs. In France, the Interministerial Committee for International Cooperation and Development reviews all funded projects and takes decisions on follow-up action. Also in the case of France, joint assessments are undertaken with national partners in bilateral projects and programs.

Fourth, the requirements for reporting by external implementing partners also vary widely. Seven of the nine surveyed countries reported funding of external evaluations of government projects and programs related to migration and development – Belgium, Denmark, Finland, France, the Netherlands, Spain, and Switzerland – although in the case of Spain and Switzerland, external evaluations were reported to take place only very occasionally. In all the responding countries (France did not respond to this particular question), operational organizations that receive government funding for projects are required to conduct their own internal monitoring, evaluation, and assessment. Standard requirements are technical and financial reports of activities undertaken, generally including quantitative information on outputs, such as the number of people served by each project. Belgium, for example, requires an activity report at the end of each semester of a project and an annual financial report. Finland applies a standard procedure to all ex ante impact assessment for projects in the development field that specifically requires an assessment of impact on gender equality, social equality, rights and opportunities for vulnerable groups, and HIV/AIDS. The Dutch government normally assesses the organizational capacity of external parties to undertake monitoring, evaluation, and impact assessment, before entering into a contract.

Challenges to Effective Impact Assessment

Previous studies have identified some of the reasons why it is difficult to assess the impact of migration policies in general and their impact on development in particular. An important reason is simply that assessing the impact of migration policies is complex. It can be difficult to establish a cause-effect relationship between the policy and its outcomes. In the case of asylum, for example, there is a continuous debate about whether it is policies in destination countries, or conditions in origin countries, or intermediary factors such as social networks and migrant smuggling that most impact on numbers (Koser 2010b). There is also usually a time lag around most data on migration and related impacts, making it difficult to demonstrate any immediate effect even where there is one.

The study by Ardittis and Laczko (2008) cited above identifies a number of additional and largely technical challenges that may also need to be overcome to improve migration policy assessment. First, many public administrators tend to hire evaluation experts, rather than migration experts, to conduct assessments of policy implementation. Second, there is usually limited interaction between public administrations and migration researchers and limited mobility between the civil service and academia. Third, NGOs are rarely involved in conducting assessments or in formal government-appointed independent advice bodies. Fourth, impact assessment reports tend to be published selectively, without any explicit criteria guiding the publication policy. Finally, there are rarely systematic rules and mechanisms to ensure that impact assessment findings can feed into policy formulation or revision.

A more conceptual challenge to effective impact assessment is identifying the intended outcomes and the indicators of success. Is the objective of migration policies that intend to impact development to help lift migrants' families out of poverty, or contribute to the development of local communities, or sustain entire states through an economic or post-conflict transition, or reduce out-migration, or some combination of all of these? Unless the objective, and how it is to be measured, is defined at the outset, then assessing the impact of a project or program or policy is difficult.

As a supplement to those already identified in earlier studies, a number of national respondents to the "Informal inquiry on policy impact assessments" also identified specific challenges related to monitoring, evaluation, and impact assessment of projects, programs, and policies on migration and development. Some of these challenges and issues were also identified during the GFMD seminar in Vienna in June 2010 and a follow-up thematic meeting on the issue in Marseille in June 2011.[2] One is that assessments are in many instances limited to the monitoring of national impacts since there exist no formally established assessment mechanisms to measure the policy impact of migration on development, including human development, elsewhere (e.g., across the border), in either the short or longer term. Second, policies that focus on the impact of migration on development are relatively recent; impact assessment outcomes are therefore sporadic and cannot be compared longitudinally. Third, and notwithstanding the complexity of assessing the longer-term impact of migration and development policies, there is a need to develop a "culture of evaluation," and governments need to recognize the importance of creating the necessary assessment mechanisms.

In addition, it was suggested that assessments should take account of the interlinkages between policies, objectives, activities, and related assessment indicators. Specifically, they should define the policy (including the challenge to be addressed and longer-term goal to be achieved), determine the objectives of the policy, set out the activities (and identify the sequence of project actions to achieve these objectives), and define the result indicators that are best suited to measure the attainment of the objectives and the expected impact of the policy. Among other tools, a logical framework approach may be useful in defining this interlinkage between policy, objectives, activities, outcomes, and impacts.

Finally, capacity building within government is required both to conduct impact assessments and to commission and review impact assessments by external evaluators. There is a serious gap between government assessment expertise and the more advanced knowledge of international agencies and expert groups. Governments and specialized agencies and experts could share their monitoring, evaluation, and assessment experiences and findings to develop "collective knowledge."

Moreover, evaluations can be expensive, an obstacle for any government, but more seriously for lower income economies. Many governments simply lack the resources and capacity to evaluate. But this applies in some respects to international organizations such as IOM. Most IOM projects include a monitoring and evaluation component, but IOM conducts relatively few impact evaluations and employs only one full-time evaluator at its headquarters in Geneva. As discussed in Marseille, there is a need to build evaluation capacities also within national migration agencies. Joint evaluations between origin and destination countries may be efficient, to achieve economies of scale, but more difficult to negotiate, given the expenses involved and

---

[2] Both events were held with GFMD governments and other actors under the auspices of the GFMD Working Group on Policy Coherence, Data and Research, and in support of the thematic programs of the Mexican and Swiss GFMD Chairs, respectively. The Marseille meeting, organized with IOM and the World Bank, was a concrete response to the recommendation made by governments at the 2010 GFMD meeting in Puerto Vallarta to better utilize data and research tools to help promote policy and institutional coherence and a "culture of evaluation" around migration initiatives.

the frequently divergent interests of either side (IOM 2010). The discussions at the GFMD meetings in Vienna and Marseille over the past 2 years reflected on the role that the GFMD could play in promoting more collaboration between states, also in terms of sharing the costs of conducting an evaluation.

Some examples of effective practice also emerged from the "Informal inquiry on policy impact assessments." The existence and experience of interdepartmental bodies to review migration and development policies, including their impact assessment, were reported on positively. The identification of specific benchmarks for evaluating the impact on human development of projects and programs (e.g., their impact on gender equality or rights and opportunities for vulnerable groups) was highlighted as allowing for more systematic reporting by external implementing partners. Joint impact assessments involving bilateral national partners were reported on positively. Another model identified for exchanging experience on impact assessments was the project "Linking emigrant communities for more development – Inventory of institutional capacities and practices," implemented within the framework of Pillar II on Migration and Development of the Dialogue on Mediterranean Transit Migration.

In the case of Mauritius, which currently runs a number of circular migration projects with Canada, while it is too early to assess the overall impact of these projects, they are closely monitored: before departure, migrants are informed of living and other conditions in Canada, followed by extensive interviews in Canada with the Mauritian workers, their employers, and their relatives in Mauritius. This has allowed for the systematic collection and analysis of information as the projects progress, including a preliminary assessment of their effectiveness. In October 2009, a major monitoring and evaluation exercise was conducted along these lines. Mauritius has also signed a circular program agreement with France, which will be implemented in the near future. This agreement also includes provisions for the economic and social reintegration of the workers back in Mauritius.

At the GFMD thematic meeting in Marseille in 2011, New Zealand showcased its "Recognized seasonal employers scheme" established in 2006 specifically to benefit low-skilled migrant workers from the Pacific Islands and employers in the horticulture and viticulture industries in New Zealand. This features a joint evaluation advisory committee, which through regular review and adjustment of the program, has been able to achieve its objective of assisting and advancing lower income workers and their families from neighboring Pacific Islands. Also the circular migration programs between Mauritius and Canada, which include monitoring and evaluation as part of the agreement, have indicated positive impacts on development.

Developing Capacity for Impact Assessment

In many less developed countries of origin, however, more basic and initial steps are required to provide even the basis for impact assessment. One is simply to recognize the linkages between migration and development, a second is better data and research on migration (Commission on International Migration Data for Development Research and Policy 2009), and a third is the need to develop a "culture of evaluation."

Poverty Reduction Strategy Papers (PRSPs)

With respect to recognizing the linkages between migration and development, it is worth mentioning briefly Poverty Reduction Strategy Papers (PRSPs), as they can be key strategy documents for providing a basis to integrate migration into development policy and practice and assessing impact.

In the last 2 years, four more countries have submitted new Poverty Reduction Strategy Papers – Nicaragua, Tajikistan (expanding on the March 2009 PRSP), Togo, and Uganda. All four make reference to migration, in ways that reflect the specific national context.

- Nicaragua's PRSP, for example, identifies emigration as a potential solution to the problems of poverty and unemployment, but equally notes the impact of the global financial crisis on rising unemployment among Nicaraguan migrants abroad, in turn resulting in a reduction in remittances. It also recognizes the need to improve border management.
- Tajikistan's PRSP also identifies emigration as a key element in reducing unemployment and generating remittances. It identifies specific obstacles to labor migration, including the lack of specific skills and insufficient financial resources; calls for more bilateral agreements on labor migration and return; and focuses on protecting the rights of Tajik migrants.
- Togo's PRSP identifies internal migration as an important reason for inter- and intra-regional imbalances and proposes the establishment of agricultural, industrial, and service "hubs" to try to reduce the pressure on internal migration, especially from rural to urban areas.
- Finally, Uganda's PRSP notes an increase in remittances between 2000 and 2007 (there is no reference to the more recent impact of the global financial crisis on remittances) and acknowledges the continued emigration of health professionals.

The inclusion of migration issues in all four of these most recent PRSPs demonstrates the continuing integration of migration into development policy and practice. The different migration issues that are addressed across the four PRSPs also demonstrate the way that migration and development are related in different ways in different national contexts. At the same time, what is missing in all four papers are specific policy proposals for enhancing the contribution of migration to development.

Mainstreaming migration into national development plans can also be an important first step toward including migration in evaluations of development projects, which tend to be more routinely assessed for their impacts on, for example, achieving the Millennium Development Goals.

Migration Profiles for Better Data and Research?

A second basic necessity is better data and research on migration, and in this regard, there has been growing attention on migration profiles, referred to here for their relevance for impact assessment. Migration Profiles were first proposed by the European Commission (EC) and initially intended as a concise statistical report gathering information on issues such as the labor market situation, unemployment rates, labor demand and supply, skills shortages by sector and occupation, skills available in the diaspora, migration flows, and financial flows including remittances (European Commission EC 2005).

The GFMD Working Group seminar in Vienna identified a number of key issues regarding the purpose, process, and character of Migration Profiles. At the seminar, governments such as Ghana and Moldova briefed participants on their concrete experiences with Migration Profiles and stressed their usefulness in fostering effective migration and development policies and systems by means of comprehensive data and enhanced government knowledge on migration trends and related development implications.

In particular, the GFMD Working Group encouraged the development of extended Migration Profiles. These incorporate a wider range of issues such as income levels, human development indicators, migrants' access to rights, "brain drain," health, education, trade, and other factors related to the impact of migration on the development of concerned countries. They also draw on a wider range of sources and involve a more elaborate process involving consultation with a variety of actors. Together these extensions can make Migration Profiles an even more effective tool for enhancing coherence between migration and development policies and providing a solid basis for impact assessment.

A series of findings and conclusions on migration profiles emerged from the Vienna seminar:

First, their content should be extended to include a comprehensive range of issues relevant in the migration and development context.

Second, they can contribute to the mainstreaming of migration factors into national development plans.

Third, migration profiles are implemented as part of national capacity building efforts to enhance government knowledge on migration and development and support governments in creating or enhancing greater interministerial coordination and institutional coherence. They are part of a wider process toward achieving more effective migration governance by providing comprehensive national data.

Fourth, and of most direct relevance to this chapter, the elaboration of migration profiles is a sustained process of activity that assesses the present and longer-term impact of migration on development and should be implemented by special interministerial task forces or similar entities to be supported by technical cooperation with specialized international agencies.

Since 2010, IOM has established basic guidelines on how to develop Migration Profiles as a first step in developing national capacity. These include a standardized Migration Profile template, a national assessment of data sources and statistical capacities, the use of local research consultants, the establishment of interministerial technical working groups, technical guidance through a Statistical Advisory Group, and National Planning Workshops. The Global Migration Group (GMG) has also been developing a core set of indicators for Migration Profiles based on estimates and data already produced by the GMG membership. And ICMPD, under the

Building Migration Partnerships (BMP) project, has created a web-based interactive map ("BMP i-Map") that also provides information on Migration Profiles and related cooperation with ICMPD.

The Swiss GFMD in 2011 moved the discussion on Migration Profiles further abroad by staging three thematic meetings, in New York, Batumi (Georgia), and Manila, on Migration Profiles. GFMD efforts at these meetings were trained on promoting the establishment and use of migration profiles and achieving some comparability between countries. In Manila, there was also a challenge to expand their usefulness to providing sufficient information on contract workers abroad to support emergency responses and possible evacuations of labor migrants in times of crisis.[3] With the assistance of IOM, the GFMD has established a Migration Profiles Repository on its website, in direct follow-up of recommendations made by governments during the 2010 Puerto Vallarta meeting, and echoed at the 2011 thematic meetings (http://www.gfmd.org/en/pfp/migration-profiles).

Fostering a Culture of Evaluation

A third fundamental requirement to provide the basis for impact assessment is the development of a "culture of evaluation," defined by the US General Accounting Office as "…a commitment to self-examination, data quality, analytic expertise, and collaborative partnerships" (US General Accounting Office GAO 2003, p. 1). It may be particularly difficult for governments to undertake policy impact assessments since in many instances they lack appropriate capacity and established assessment mechanisms. As migration-related policies are often spread across or funded by different government departments, a further challenge is coordinating monitoring, evaluation, and impact assessment.

Governments also need to be prepared to accept mixed results from impact assessments, including low impact, unintended consequences, and negative outcomes, and this is probably especially difficult in politically sensitive fields such as migration. In this regard, the GFMD/IOM/World Bank workshop on "Managing migration for development: Policymaking, assessment, and evaluation" identified a "fear factor" that often pertains to monitoring and evaluation whereby governments see evaluation as a threat rather than an opportunity to improve the effectiveness of their interventions (Chappell and Laczko 2011).

Important questions to ask in this regard are: How can a national "culture of evaluation" be fostered and strengthened? Which actors should "champion" this cause? In many cases, the answer to such questions may be the establishment of a specific

[3] See the meetings: "Migration profiles: developing evidence-based migration and development policies" in New York, USA, June 30, 2011, "Migration profiles: lessons learned" in Batumi, Georgia, July 12–13, 2011, and "Migration profiles as a tool for informed policy making, integration and reintegration and emergency response" in Manila, Philippines, October 20–21.

organizational framework, as well as legislation requiring impact assessment for all publicly funded projects. The Marseille workshop also concluded that the recent events of the Arab Spring may provide a unique opportunity to energize new decision-makers as well as development practitioners and international experts.

Conclusions

At the 2009 Athens meeting of the Global Forum on Migration and Development, participants called for more research aimed at evaluating a range of programs and policies, including assessments of their impacts on the development of communities, families, and migrants themselves in both origin and destination countries. This chapter has reported on some initial findings of limited research conducted by the GFMD ad hoc Working Group on Policy Coherence, Data, and Research in response, supplemented by findings from earlier studies on impact assessment for migration policy, and follow-up actions by the GFMD in 2011.

An immediate conclusion is that far more research is required: a more extensive survey should identify further examples of effective practice on impact assessment as well as the challenges and capacity requirements involved. At the national level, the best developed and most effective policy impact assessments often take place in sectors other than migration, for example, in public health, environmental impact, and development including micro-finance, and there may be lessons to learn across national sectors through cross-fertilization. An interdepartmental body might be best suited to oversee monitoring, evaluation, and impact assessment across government.

At the international level, it may be useful for the GFMD to follow the outcomes of the joint facility on migration management for human development launched by the World Bank and IOM in 2011, hosted by the multi-partner platform of the Marseille Center for Mediterranean Integration (CMI) by the fall of 2011.[4]

Besides identifying a need for more research, this chapter has also identified a series of concrete recommendations. Fundamental to impact assessment are the recognition of the migration-development nexus in policy and practice, good data and information on migration and development, and a culture of evaluation. Beyond such fundamentals, national governments need to identify the most effective way to undertake impact assessments of migration and development policies, which will probably be determined by national priorities.

Whether undertaken by government staff or external evaluators, effective tools and approaches for impact assessments are required, which should include the identification of clear policy objectives, the definition of appropriate indicators linked to development, and agreement on the intended beneficiaries. A standard system is

[4] See the Summary Report of the GFMD 2011 thematic meeting in Marseille, "Managing migration for development: Policymaking, assessment and evaluation," June 13–15, 2011.

also needed for requiring external implementing partners to provide impact assessments. Greater cooperation with international organizations or expert groups can assist governments develop better methodologies for impact assessment.

There is clearly scope for a "clearing house" that allows governments to learn from one another's experience with impact assessment of migration and development policies, including more systematic exchanges with relevant international agencies (Ardittis and Laczko 2008). Especially as migration and development policies almost always impact several countries, joint impact assessments on a bilateral or regional basis would allow the full developmental impacts of migration policies in both origin and destination countries to be assessed, while simultaneously promoting dialogue and potentially also contributing to capacity development in less developed countries (Chappell and Laczko 2011). The first steps taken by the GFMD in setting up a Migration Profiles Repository may be a useful first step toward improving future data collection and assessment of policies and their impacts on development.

References

Ardittis, S., & Laczko, F. (2008, April 2008). *How are the costs and impacts of migration policies evaluated?* (Migration Information Source). Retrieved from http://www.migrationinformation.org/Feature/display.cfm?id=677

Chappell, L., & Laczko, F. (2011). *What are we really achieving? Building an evaluation culture in migration and development* (Migration Information Source). Washington DC: Migration Policy Institute.

Commission on International Migration Data for Development Research and Policy. (2009). *Migrants count: Five steps towards better migration data.* Washington DC: Center for Global Development.

European Centre for Health Policy (ECHP). (1999). *Health impact assessment: Main concepts and suggested approach.* Brussels: ECHP.

European Commission (EC). (2005). *Communication on migration and development.* Brussels: EC.

International Organization for Migration (IOM). (2010). *World Migration Report 2010: The future of international migration – Building capacities for change.* Geneva: IOM.

Koser, K. (2009). *Responding to the impact of the global financial crisis on international migration: Lessons learned from previous financial crises* (IOM Migration Research Series, Vol. 37). Geneva: IOM.

Koser, K. (2010a, September). *Assessing the impact of migration on economic and social development.* Background Paper prepared for Roundtable 3.1 of the Global Forum on Migration and Development.

Koser, K. (2010b, December). *Responding to boat arrivals in Australia: Time for a reality check* (Lowy Institute Analysis). Sydney: Lowy Institute.

US General Accounting Office (GAO). (2003). *Program evaluation: An evaluation culture and collaborative partnerships help build agency capacity.* Washington DC: GAO.

Chapter 12
Textbox 5: Measuring the Household Effects of Temporary Overseas Work: A Unique New Study in India

Michael Clemens

Millions of workers from developing countries migrate overseas each year for temporary, high-wage work. Indians, Pakistanis, and other foreigners constitute the majority of the population in countries like the United Arab Emirates (UAE) and Qatar. These flows have become large enough to consider them as a force shaping global development.

But how does migration shape the lives of guest workers and their families? Large numbers of workers seek jobs abroad despite frequent reports that many are poorly informed about the wages and working conditions they will have abroad. And while workers send large volumes of money home, reservations are often expressed about the development impact of those remittances. Do they increase investment or reduce economic activity at home by encouraging remittance recipients to work less?

A new study has been undertaken with the support of the UAE Ministry of Labor and the John D. and Catherine T. MacArthur Foundation to measure the effects of temporary wage work in the UAE on Indian construction workers and their families back home. The study uses a so-called 'natural experiment': the large shocks to migration from the construction crisis in the UAE.

The experiment arose at the end of August 2008, when the global financial crisis led to the sudden freezing or cancelation of large numbers of UAE construction projects, particularly in Dubai. International finance dried up amid the wave of instability (the same wave that would destroy the American investment bank Lehman Brothers 2 weeks later), and domestic finance dried

M. Clemens (✉)
Center for Global Development, Washington DC, USA
e-mail: mclemens@cgdev.org.

I. Omelaniuk (ed.), *Global Perspectives on Migration and Development: GFMD Puerto Vallarta and Beyond*, Global Migration Issues 1, DOI 10.1007/978-94-007-4110-2_12,

up as the price of the UAE's chief export – petroleum – plummeted 60% in 2 months. Hundreds of construction projects halted, some for months and others for years.

Many UAE construction firms canceled work contracts for workers from India and elsewhere, including hires already in process for a UAE visa. Small differences in the date that an Indian worker applied for the UAE job are associated with large differences in the probability that he arrived in the UAE and stayed to work.

But small differences in the date that he applied for his UAE job are unlikely to be associated with large differences in the observable or unobservable traits of that worker and his family. Because the variation in migration behavior due to the crisis had nothing to do with individual households' characteristics, differences between households with and without a migrant *due to the crisis* can be confidently attributed to the pure effects of migration itself.

The next step was to acquire data on the Indian workers and their families. The study links three different sources of data, beginning with a list of several thousand Indian workers who were selected for construction jobs in the UAE through one major multinational construction firm between mid-2008 and mid-2009. A survey firm visited the home addresses of those workers in late 2011 to administer a detailed household questionnaire on migration behavior, finances, and other traits. Administrative records from the UAE Ministry of Labor were used to characterize the work and compensation of those who had gone to work in the UAE. All three datasets are matched by individual identifiers.

The survey work was completed in November 2011, and the initial results of the study suggest that migration to the UAE increases employment and wages for otherwise identical Indian workers, combining to raise their incomes by 250–350%. Migration does not cause declines in labor force participation or earnings by other workers in the migrants' household. Migration to the UAE causes the migrant's household in India to be more likely to receive income from a home-based business.

Chapter 13
Regional and Inter-regional Processes: Advancing the Discourse and Action on Migration and Development

Maureen Achieng

Viewing the RCPs and Inter-regional Fora Through the GFMD

From its beginnings in 2007, the Global Forum on Migration and Development (GFMD) has focused on some aspect or other of the role, purpose, and achievements of Regional Consultative Processes on Migration (RCPs) and Inter-regional Fora (IRF). This is not surprising given the increasing importance of RCPs as multilateral fora for dialogue and cooperation on migration issues, in some instances specifically focused on the mutual development benefits, and the fact that the GFMD itself is modeled on the RCP format. This discussion has evolved over the course of the GFMD process.[1]

In 2007, the GFMD in Brussels sought to explore the impact that RCPs had on advancing cooperation between states on migration and development issues (GFMD 2007a). The background paper (GFMD 2007b) that guided the discussion on this topic observed that RCPs were not in a position to engage on issues relating to migration and development because officials participating in these processes were predominantly from ministries responsible for immigration. It suggested that the potential role RCPs could play in enhancing the contribution of migration to development remained underexploited.

The principal RCPs were generally seen as falling into three categories:

1. Those for whom migration and development is a thematic priority
2. Those that include migration and development in their agenda on an ad hoc basis
3. Those not concerned with the issue in any shape or form

[1] See the Roundtable 3 discussions of Brussels, Belgium (2007) through to Puerto Vallarta, Mexico (2010). Retrieved from http://www.gfmd.org. Accessed 21 Nov 2011.

M. Achieng (✉)
International Partnerships Division, International Cooperation and Partnerships Department, International Organization for Migration, Geneva, Switzerland
e-mail: Machieng@iom.int

I. Omelaniuk (ed.), *Global Perspectives on Migration and Development: GFMD Puerto Vallarta and Beyond*, Global Migration Issues 1, DOI 10.1007/978-94-007-4110-2_13,

Various factors were identified that influence the inclusion of migration and development in the agendas of RCPs. First, how an RCP is funded influences to a great extent what issues it chooses to focus on, beyond or in addition to those that triggered its establishment. As donor funding will generally tend to be earmarked, the focus issues will invariably also reflect the priorities of the donors. This could in some instances broaden the RCP agenda to include development issues linked to migration. Conversely, donor funding priorities could undermine the possibility that these discussions are enlarged to include development factors.

Second, the trend toward broader ministerial representation at these meetings has the potential of ensuring a stronger development focus. Also, and perhaps not surprisingly, those RCPs with a strong representation of developing countries have tended to integrate development concerns into their discussions. Even RCPs comprised wholly of developed countries, such as the Inter-governmental Consultations on Migration, Asylum and Refugees (IGC), sometimes focus on development issues, given the growing concern to ensure policy coherence in the field of migration, which is now widely understood to demand a multisectoral approach (Ibid., p. 7).

Finally, these first GFMD deliberations called for greater cross-fertilization between RCPs through the sharing of information and good practices not only within regions but also across regions (Ibid.). Establishing formal links between RCPs and regional organizations[2] and trade regimes was considered to be positive by some RCPs, which felt that the association gave them more leverage and could help ensure follow-up on recommendations. Additionally, proposals for possible approaches to integrating development into RCP deliberations were considered, as was the role of the regional processes vis-à-vis the GFMD.

Broadly, while recognizing that RCPs are state-led, the first GFMD urged its members to work toward more systematic information sharing on their respective activities and proposed that an assessment be conducted to determine the impact that RCPs have had to date on migration governance. Member states should strive for greater policy coherence at national level, ensure more predictable donor support, establish new RCPs in regions where there were none, and promote linkages between RCPs and other regional fora dealing with migration. RCPs play a complementary role to the GFMD, with the former providing "a testing and dissemination ground for new ideas" (Ibid., p. 11) and the latter producing new ideas in relevant areas, thereby creating a "sustainable two-way information flow between RCPs and the GFMD" (Ibid.). It was felt that, over time, this could enhance the role of the regional processes, also as building blocks for greater international cooperation on issues of migration and development.

The second GFMD in Manila, 2008, cast a broader focus on the issue, looking at how the full range of RCPs, inter-regional fora (IRF), regional organizations, and

---

[2] The Inter-governmental Authority on Development RCP (IGAD-RCP) which is formally linked to the six-member organization of East African countries that makes up the Inter-governmental Authority on Development (IGAD) is a good example of this. Similarly, membership of the Southern African Development Community (SADC) is aligned to that of the Migration Dialogue for Southern Africa (MIDSA).

economic integration processes interfaced with migration and development (GFMD 2008b). Many of the earlier points about the regional processes being state-led and created specifically to focus on migration issues of immediate concern to the member states, rather than on the migration-development nexus, were reinforced. Regional and sub-regional organizations that add migration to their agendas were recognized as providing additional venues for inter-state cooperation on migration. Notably, several of the IRF, straddling major regions of migrant origin and destination, were thought to focus more on solutions that touched on both migration and development.

The Manila discussions sought to advance understanding of all three categories of fora that focus on migration exclusively or in part. The contribution of these fora to enhancing inter-state cooperation on migration was acknowledged, while recognizing the common challenges they faced in working toward the realization of a predetermined set of objectives. Manila took stock of significant developments that had taken place since the Brussels GFMD, such as the establishment of new RCPs, and consultations that had taken place between RCPs as well as between them and other IRF across the five major world regions,[3] while noting that important policy challenges remain.

There was an endeavor to highlight regional and inter-regional practices that have proven effective in promoting inter-state dialogue on migration, especially at the development interface. The Manila GFMD in 2008 outlined the key policy challenges and echoed some conclusions from previous GFMD discussions on this theme, such as the need to achieve coherence among all of these fora, ensure effective implementation of outcomes, and conduct impact assessments (GFMD 2008a).

The Athens GFMD in 2009 noted some key similarities between RCPs and IRF: All are state-led, informal, and non-binding dialogues on migration and operate outside any institutional framework. In contrast to RCPs and IRF, the GFMD functioned at the global level, grappling with issues that cut across regional processes and their location-specific interests (GFMD 2009a, pp. 36–39). Building on previous GFMD deliberations, the Athens GFMD explored how the GFMD could learn in practical ways from successful regional and inter-regional processes, and vice versa. It looked at possible roles for regional and inter-regional processes in implementing GFMD recommendations and follow-up actions (GFMD 2009b). Finally, the Athens GFMD reviewed the important developments that had taken place since the previous GFMD meetings, some in direct response to actions recommended at previous GFMD roundtable discussions on this theme (Ibid.).

The parallel civil society discussions on this theme reflected a wide range of views on the possibility of linking RCPs and IRF to the GFMD process. The outcomes of these discussions suggested that civil society organizations (CSOs) could play a positive role in these state-led and state-owned dialogues. Similarly, RCPs were thought to be potentially valuable to the work of CSOs, as they could increase

[3] As follows: the Americas, Asia-Pacific, Africa, the Middle East, and Europe.

visibility and dialogue on migration and related issues of concern to these organizations. A predominant view among CSO representatives was that the informal nature of RCP deliberations and their non-binding outcomes greatly diminished their potential to have any real impact.

By way of example, the Athens CSO deliberations on this topic suggested there was a need to integrate a rights-based approach into RCP exchanges since this was an issue where the on-the-ground expertise of CSOs could add value (GFMD 2009c, pp. 6–7). CSO-proposed objectives for the regional processes included improved information exchange and coordination among them, capacity building, and evidence-based policy making, in addition to greater investment in the processes. These were thought to be concrete actions that could benefit countries of origin, transit, and destination, as well as migrants and their families, particularly if civil society were brought into the dialogue.

Overall, GFMD deliberations on this topic up to this point suggested an improved understanding of the role played by RCPs and IRF with a migration or migration-development focus. These first three GFMD discussions on the subject provided an opportunity for migration policymakers and practitioners to enhance their understanding of how the existence and actions of RCPs and IRF are advancing knowledge and action on the migration and development connections. These sustained GFMD deliberations on the role of RCPs and IRF also have the potential in time to influence the evolution of agendas of RCPs and IRF.

GFMD Puerto Vallarta: A New Paradigm on the Role of RCPs and IRF?

GFMD Puerto Vallarta in 2010 took stock of deliberations to date under the topic "How can Regional Consultative Processes On Migration and Inter-Regional Fora best include the migration and development nexus" (GFMD 2010a).

The background paper for this discussion in Puerto Vallarta called attention to the emergence of a new desire by states within the same region and across regions, to discuss migration issues in depoliticized fora. Drawing on outcomes of previous GFMD meetings, and informed by the evolution of migration and development discussions within their respective regions, several RCPs and IRF were stepping up the dialogue on themes more closely related to migration and development than previously. Several RCPs held pre-GFMD consultations to plan for their participation at the GFMD, an indication of the extent to which some countries/regions were keen to showcase their migration and development initiatives. Some regional and inter-regional dialogues had broadened their substantive and thematic coverage of these two fields.

To advance the discussion from what previous GFMD fora had accomplished, GFMD Puerto Vallarta suggested the need for a new approach to how RCPs and IRF, and their contribution to the discourse on the migration and development nexus, are viewed (GFMD 2010b, pp. 40–43).

Enhanced capacity[4] was considered key to fostering the necessary policy and institutional coherence through a "whole-of-government" approach that would facilitate the inclusion of development in migration policy discussions (IOM 2003). Improved understanding of migration and its linkages with other domains would probably result in a greater focus by RCPs on the developmental impact of migration, for example, the potential contribution of migration to advancing the Millennium Development Goals (MDGs) (e.g., through social and financial remittances), the benefits that could accrue to states from dialogue and cooperation with diaspora groups, and measures to address brain drain and brain circulation in the context of development).

Lack of national capacity was felt to be a critical concern in this debate, as weak migration governance at national level invariably translates into less capacity at the regional or inter-regional levels. Thus, ensuring that the migration-development relationship is taken into account in the deliberations of RCPs and IRF may well be a question of enhancing the capacities of participating countries to actually do so. This could then also extend to the capacities of the RCPs and IRF they participate in.

Since states set the agendas of RCPs and IRF, the stronger and better established the member states are in this area of governance, the more they may be able to evolve and adapt to changing migration concerns in the RCPs or IRF they belong to. States and the regional and global processes they participate in can be said to feed off each other. While states establish the agendas of the processes, their own national policies and practices could in time be influenced by the broader directions of the regional and global processes they participate in. There are thus strategic advantages to bringing the migration and development connection more firmly into the agendas of RCPs and IRF, where it had not been a declared thematic priority at the outset (GFMD 2010c).

The background paper for Puerto Vallarta suggested the need for a new paradigm to view and assess the potential or actual contribution of RCPs and IRF to the migration and development debate (GFMD 2010a, pp. 5–6). An assessment of RCPs commissioned by IOM (Hansen 2010) suggested that their impacts on migration governance needs to be understood in a fundamentally new way: They should not be evaluated solely for their impacts at national level but also in terms of their complementary role vis-à-vis global processes such as the GFMD. Their contributions are wide-ranging and complex and tend toward outcomes that enhance trust

[4] The IOM World Migration Report (IOM 2010, p. 7), defines capacity building as the process of strengthening the knowledge, abilities, skills, resources, structures, and processes that states and institutions need in order to achieve their goals effectively and sustainably and to adapt to change. More specifically, capacity building in migration management includes as key components the more timely and accurate migration and labor market data; assistance in defining national migration policy goals and priorities, training of migration officials; development of an effective and equitable legal framework, coherent administrative structures; consultations mechanisms between government and other national stakeholders; and international cooperation. Retrieved from http://publications.iom.int/bookstore/free/WMR_2010_ENGLISH.pdf. Accessed 21 Nov 2011.

between states and improve understanding on migration issues. They create consensus and build states' capacity in this domain, which in turn de facto influences changes in their institutional arrangements, laws, policies, and practice at the national and, and at times, the regional level.

It was argued in Puerto Vallarta that since all RCPs and IRF focus on some aspect of migration, they contribute by their very existence to new thinking and new ways of addressing migration. Regular participation in an RCP will often result in improved communication and coordination among government agencies at the national level, and a common understanding and definition of key concepts, leading in turn to enhanced capacity and policy coherence at national level. This can only lead to improved migration governance, in itself a critical factor for linking migration and development.

For example, the Bali Process[5] has convened regular meetings at the ministerial, senior officials, ad hoc working group, and experts levels in the 9 years of its existence and made significant inroads in the development of regional protocols that have translated into national level interventions to respond to the challenges of trafficking in persons and human smuggling, and associated with this are positive human development outcomes for Bali Process countries.[6]

Equally illustrative is the Puebla Process,[7] which ensures both technical and political inputs into its deliberations through the two-tier structure it has adopted. All its meetings produce reports outlining a set of political, non-binding recommendations. The Regional Consultations Group on Migration (RCGM) and the RCM-established liaison officer networks (LON) are then invited to report back on follow-up activities to the vice-ministerial meeting (VMM). Experience has shown that while none of these recommendations are binding on the Puebla Process member states, several of them are being implemented in a manner that could be seen as moving toward de facto regional policy harmonization, even if this is not their original objective.

If we accept that RCPs and IRF, whatever aspect of migration they choose to focus on, are positively affecting the migration and development relationship, then perhaps the discussion should focus less on how these fora can take better account of the migration and development nexus, and more on how they could be strengthened for the role they are performing or have the potential to perform. This position is critical given the continued absence of a global approach to migration supported by appropriate institutional apparatus.

The civil society discussion paper on regional and inter-regional processes for the Puerto Vallarta GFMD argued that many RCPs and IRF already deal with the migration and development nexus (Newland 2010). The flexibility that RCPs have to establish their own agendas was felt to answer the question of how RCPs and IRF

[5] The Bali Conference on People Smuggling, Trafficking in Persons and Related Transnational Crime.

[6] See the Bali Process website, available at http://www.baliprocess.net/. Accessed 21 Nov 2011.

[7] Also known as the Regional Conference on Migration (RCM).

could incorporate the migration and development nexus. Many delegates pointed to the need for improved RCP capacity through established secretariats with longer-term staff and more regular meetings, given the common experience of irregular meetings leading to a loss of momentum. Another key obstacle was felt to be the underrepresentation of development agencies relative to foreign policy, migration, labor, and justice/home affairs on many delegations to RCP meetings (Ibid., p. 6).

Overall, the civil society discussion paper advanced the view that while the RCP framework may not be the best-suited mechanism through which to tackle the relationship between migration and development, the existence of RCPs and IRF could over time result in more migrant-friendly policies (Ibid., p. 8). However, echoing the sentiments of the RCP assessment study, it noted that "major shifts in migration and development policy" are not likely, and building inclusive policy networks to facilitate communication across borders, bureaucracies, and social barriers would at this point seem to be more within the reach of RCPs and IRF (Ibid.).

Both government and civil society discussions in Puerto Vallarta on the role of RCPs and IRF were animated and drew in a large spectrum of views. However, it would be difficult and premature to determine the extent to which there is improved understanding by migration policymakers and practitioners following Puerto Vallarta and the earlier GFMD discussions on this subject. Nonetheless, it seems reasonable to say that the Puerto Vallarta discussions on the role of RCPs and IRF may have set the discourse in a new direction, particularly on the government side.

In broad terms, the government discussions in Puerto Vallarta underscored the importance of leadership of these regional and inter-regional fora, as well as the need for stronger inter-governmental coordination structures. Despite some calls for binding approaches and outcomes of RCP meetings, many states continued to support the non-binding nature of the regional and inter-regional processes. Others argued that if they are left as non-binding dialogue processes, their impact will remain limited. The decision about whether or not to move toward greater formality by ensuring binding outcomes is clearly for the individual states/RCPs/IRF to determine and cannot be imposed on them. Several governments reaffirmed the need for greater cross-fertilization between RCPs through meetings that bring them together across regions as well as globally, such as those that take place in the context of IRF and global meetings of RCPs (GFMD 2010b, pp. 41–42), the third and most recent of such meetings having taken place in Africa in 2011, precisely on the issue of capacity enhancement of RCPs.[8]

The fact that migration is not always prominent on national agendas was seen as impeding progress toward greater international collaboration on migration management. There were also calls in Puerto Vallarta for better data collection, disaggregated by sex, and increased sharing of data by states, with migration profiles offered as a possible tool to bolster evidence-based migration policy planning and practice. On the

[8] The Third Global Meeting of Chairs and Secretariats of Regional Consultative Processes on Migration (RCPs) was held in Gaborone, October 25–26, 2011.

question of strengthening the potentially mutually reinforcing relationship between RCPs/IRF and the GFMD, several states expressed the view that as RCPs and IRF are independent state-led fora, any attempt at institutionalizing linkages between them would be ill-advised and would in any case be premature at this point.

A minority of government participants in Puerto Vallarta highlighted the need for some civil society participation in the deliberations of RCPs and IRF. The introduction by the Mexican GFMD Chair of a Common Space segment in the GFMD – with the aim of creating an opportunity for informal, open, and solution-orientated dialogue between government and non-governmental actors participating in the GFMD – was hailed by many as a possible model for future interaction between regional and global fora for dialogue on migration (Ibid., p. 13).

Lastly, several government participants expressed the view that there was no urgency to create RCPs in regions that did not have them, as this may result in state-led processes that remain heavily dependent on external facilitators (Ibid., p. 43).

On the civil society side, there was a somewhat unequal understanding of RCPs and IRF, resulting in a discussion that ranged from perceiving them as highly ineffective due to their non-binding and informal nature to not knowing of their existence. The issue of civil society participation was also examined in the case of RCPs that permitted it, and it was argued that direct participation was not the only way for civil society actors to affect outcomes. Other available options included critiques of RCP plans of action, presentation of research findings, or expert analyses of topics that were before RCPs. These avenues needed to be explored and utilized wherever possible.

Despite the wide-ranging discussions in Puerto Vallarta, some questions remained or arose. For instance, how does this discussion on the (potential) role of RCPs and IRF in addressing the migration-development nexus advance international cooperation on migration? What is the way forward, given the consensus that unilateral approaches to migration are insufficient to address migration challenges and the development opportunities that it offers?

As with RCPs and IRF, the GFMD provides a platform through which states could achieve shared understanding of common migration concepts and gradually move toward enhanced cooperation, in some instances possibly some policy convergence over time. Today, states mostly recognize the inadequacy of unilateral approaches and are doing more to go beyond bilateral approaches to migration management toward enhanced regional cooperation. Notwithstanding the persistent calls for a global migration governance regime, international cooperation and action on migration still remains limited. There is a need to look at what, in the absence of a system that facilitates global cooperation and action on migration, the regional efforts are achieving.

The following section examines what some of the regional and inter-regional processes and fora are focused on, and how these discussions may be helping (or not) to strengthen and leverage the migration and development linkages.

RCPs and IRF: Advancing the Migration and Development Linkage

Since RCPs and IRF are not, and do not seek to be, formal governing entities, they cannot be evaluated on the attainment of objectives they have not set out to achieve. Also, since their major function is a complementary rather than determinative one, they generally seek to enhance cooperation on migration issues among their own constituencies by, for example, building trust, increasing understanding of migration issues, working toward consensus, and enhancing capacity of their members (Hansen 2010, p. 43). Importantly, the brief experience of regional dialogues seems to suggest that while RCPs and IRF do not seek to achieve policy convergence, the outcomes of some of their actions de facto go in the direction of policy convergence. This suggests that RCPs' strength lies in their "soft power," that is, information and persuasion rather than tough legislation and sanctions.

The only independent assessment of RCPs to date (Hansen 2010) indicates that the value of RCPs is in their ability to influence migration governance and that largely depends on how well they function and the extent to which they remain state-led and owned. Since today most countries in almost all regions belong to or participate in an RCP,[9] it may be rewarding to examine the impact, if any, that membership of an RCP or IRF has on those countries' and regions' migration policies and legislation. However, it would be difficult to assess what proportion of global migration flows are being influenced, directly or indirectly, by the actions or deliberations of these regional and inter-regional entities. It may be more reasonable to conclude that given the widespread and active nature of these regional and inter-regional processes, there is some potential to impact global migration flows. The extent of this remains to be established.

Based on the preceding discussion, and given the limitations of any attempt to assess the impact of RCPs and IRF on migration governance, important questions arise for policymakers in this field that require more serious reflection than is possible in this chapter. For instance, how in concrete terms are RCPs and IRF influencing migration and development governance at the national level? And, by doing so, how are they enhancing international cooperation on migration and development issues within and between regions?

These issues merit more serious consideration than the academic attention to this subject to date. The available literature suggests that there are currently three broad views on the question of the effectiveness of RCPs (Koehler 2011):

1. RCPs may actually be contributing to regional "fortress" building with regard to migration, reinforced by their links to regional and sub-regional economic communities.

[9] This number is even lower when one considers the number of countries that are part of an IRF. (An estimated 111 of these countries are part of an IRF.) Missing from RCPs are 23 countries in the Caribbean and in Central Africa. Whether and how regularly they have participated in meetings of the IRF in question was, however, more difficult to determine.

2. RCPs favor restrictive legal and security approaches to migration at the expense of human rights approaches, aided by the fact that their deliberations are by and large shielded from public scrutiny.
3. Contrary to prevailing views, the very existence of RCPs and similarly structured fora is more likely to give rise to global policy convergence through "policy transfer" than to regionally divergent and conflicting policies on migration.

Clearly, any attempt at assessing the contribution of RCPs and IRF to migration governance, especially as it links with development, reaches beyond the reflections of this chapter. Nonetheless, it is worth highlighting several initiatives that have resulted from the discussions of RCPs or IRF, which demonstrate how RCPs and IRF are influencing migration governance in general (however, with no demonstrable migration and development linkage):

(a) Since its establishment in 1996, the Regional Conference on Migration (Puebla Process)[10] has promoted dialogue and cooperation among its members on issues related to migration policy and management, human rights of migrants, and migration and development. In recent years, concrete actions have been taken to strengthen consular services for migrants of Puebla Process countries and to enhance close cooperation between them. More formal networks have been established between officials from the 11 Puebla Process countries responsible for tackling human trafficking. With the increase in extra-regional irregular migrants from other countries in the region, the Puebla Process has established a fund to support their voluntary repatriation.
(b) The Inter-Governmental Authority on Development – Regional Consultative Process on Migration (IGAD-RCP)[11] reached the decision at its inaugural Ministerial Council meeting in November 2008 to develop a regional migration policy framework to enhance collaboration on migration management among its members. Among others, this framework document lays out initial elements for a free movement protocol for citizens of the IGAD-RCP space. A senior officials meeting in October 2010 took the decision to establish Regional Migration Coordination Committees (RMCCs) comprising Directors of Immigration of the IGAD-RCP countries, as well as a Steering Committee of officials mostly from departments of immigration, labor, national security, refugees, and foreign affairs (IGAD-RCP 2010, pp. 2, 6).
(c) The Ministerial Consultation on Overseas Employment and Contractual Labor for Countries of Origin in Asia (Colombo Process 2010) provides a forum for Asian labor countries of origin to share experiences and practices to help individual countries better safeguard the well-being of their labor migrants.

[10] See Regional Conference on Migration Virtual Secretariat. Available at: http://www.rcmvs.org/. Accessed 21 Nov 2011.

[11] See Inter-Governmental Authority on Development – Regional Consultative Process on Migration. Available at: http://www.iom.int/jahia/Jahia/policy-research/regional-consultative-processes/rcps-by-region/igad-rcp. Accessed 21 Nov 2011.

Its members have worked individually, in part helped by the knowledge acquired through their joint deliberations, to establish labor market research units within their ministries of labor and gain a more accurate understanding of labor markets in key countries of destination. Several have also established Migrant Resource Centers (MRCs) and carried out awareness-raising campaigns with a view to making available accurate information to prospective labor migrants. More recently, at the fourth Ministerial Meeting of the Colombo Process in April 2011 (Colombo Process 2010), governments agreed to work on a more collective basis to enhance their bargaining power vis-à-vis key countries of destination of their labor migrants.

(d) One concrete initiative being considered is a joint handbook to guide Colombo Process Member States in situations of crisis that impact migrant workers, such as recently witnessed in the Middle East and North African (MENA), where tens of thousands of Asian labor migrants, mostly from developing countries, were stranded and in need of repatriation assistance. The handbook will look at the responsibilities of all parties: sending governments, receiving governments, employers in the country of destination, and the migrants themselves. The handbook should provide a basis for sending countries to amend their national legislation, as needed, and ensure that the necessary provisions are built into the work of recruitment agencies in the countries of origin.

(e) The Inter-Governmental Consultations on Migration, Asylum and Refugees (IGC),[12] the first RCP, established in 1985, is a forum for policy debate and information exchange on the management of international migration toward a more harmonized approach to migration, asylum, and refugees issues. The IGC has adapted to changing migration realities, shifting from its initial focus on asylum and protection to immigration and integration issues in later years. Recognizing the increase in migration resulting from man-made and natural disaster crisis situations, it recently dealt with the question of humanitarian responses to crises with migration consequences at its full round meeting of May 2011. The IGC has also published a report on asylum procedures in IGC participating states (an update is in the pipeline) and is working on a variety of other issues such as the balance between attracting foreign workers and students and preventing abuse and fraud, integration indicators, and the use of technology in asylum and immigration procedures.

(f) Since its establishment in 2000, the Migration Dialogue for Southern Africa (MIDSA)[13] has promoted dialogue among its member states with a view to enhancing institutional capacities. Part of its broader agenda is to cooperate with the Southern African Development Community (SADC) region countries on migration. At its first Ministerial Meeting in October 2010, MIDSA ministers

[12] See Hallam and associates (2005); Also available at http://www.igc.ch/. Accessed November 21, 2011.

[13] See Migration Dialogue for Southern Africa website, available at: http://www.migrationdialogue.org/midsa/. Accessed 21 Nov 2011.

reached a consensus to recommend that the 2005 SADC Protocol on the Facilitation of Movement of Persons be ratified by all SADC Member States (MIDSA 2010, pp. 19, 23).

(g) In 2002, the Bali Process on People Smuggling, Trafficking in Persons and Related Transnational Crime (Bali Process)[14] was established to strengthen regional policy and law enforcement cooperation to combat human trafficking and migrant smuggling and develop regional responses to irregular migration, including sustainable solutions for refugees. In response to the upsurge of recent years in asylum seekers in many of its participating states, the Ministerial Meeting of March 2011 discussed options for closer cooperation in processing asylum claimants. Particular consideration was given to setting up a Regional Framework for Cooperation that could ensure a more comprehensive and coherent approach to the management of irregular migration across the region.

(h) The South American Conference on Migration (SACM)[15] set up in 1999 promotes cooperation and information exchange among its members on a range of issues including development, diaspora matters, rights of migrants, integration, trafficking, and smuggling. In a 2010 meeting in Cochabamba, Plurinational State of Bolivia, the SACM adopted a South American Human Development Plan on Migration, a framework document comprising a number of initiatives for action, whose implementation would be subject to funding availability. Member states also issued a Declaration on Migration Principles and General Guidelines of the South American Conference on Migration, reaffirming their commitment, inter alia, to the protection of the human rights of migrants. At its sixth inter-sessional meeting in Santiago in 2011, SACM member states agreed to initiate a process of information gathering on their experiences of the impact of climate change on migration. A global document would be developed outlining a regional position on the issue of climate change and its impacts on migration, for presentation at the forthcoming XI SACM meeting in Brasilia.

(i) The Ministerial Consultations on Overseas Employment and Contractual Labour for Countries of Origin and Destination in Asia (Abu Dhabi Dialogue),[16] since its inception in 2008, seeks to identify the roles and responsibilities of government and private actors at all stages of the contractual work cycle and to elaborate concrete partnership projects. As it embarks in the near future on developing a regional multilateral framework on temporary contractual labor mobility, the Abu Dhabi Dialogue will pilot some innovative practices that could eventually serve as a model for other destination countries in this region with regard to managing overseas contract workers in line with internationally accepted standards of protection of migrant workers.

[14] See Bali Process website, available at: http://www.baliprocess.net. Accessed 21 Nov 2011.

[15] See OIM Conosur website, available at: http://www.oimconosur.org/varios/index.php?url=conferencia. Accessed 21 Nov 2011.

[16] See Colombo Process: Abu Dhabi Dialogue website, Retrieved from: http://www.colomboprocess.org/index.php?option=com_content&view=article&id=46&Itemid=29. Accessed 21 Nov 2011.

(j) Acknowledging the need for the European Union (EU) to expand its engagement with Africa in the field of migration, the EU-Africa Dialogue provides a framework for dialogue and ultimately the implementation of various initiatives. The Migration, Mobility and Employment Partnership, implemented by ICMPD, FIAAP, and IDEP in collaboration with the African Union (AU), seeks to foster dialogue and partnerships for the future cooperation between Africa and Europe in existing and new areas across the broad sociopolitical and economic spectrums. It was agreed in 2010 to build capacities to enhance the development impact of remittances, a project intended to benefit AU Member States, remittance senders and recipients, and other stakeholders. Concrete strategies and operational instruments are being devised to use remittances as development tools for poverty reduction.[17]

(k) Another offshoot of the EU-Africa dialogue, the East African Migration Route Programme (EAMR) on "Building co-operation, information sharing and developing joint practical initiatives amongst countries of origin, transit and destination," derives from the action guidelines suggested in the Commission-proposed priority actions for improving migration management adopted by the European Council in its conclusions of December 2005 (Council of the European Union 2005). This initiative is intended to complement those undertaken by France and Spain in West Africa in recent years.

(l) The Asia-EU dialogue[18] on labor migration, born in 2008 out of the Colombo Process and an EU-funded project to enhance cooperation with Asian countries of origin of migrants to the EU, seeks to ensure that fuller account is taken of migration trends between Asia and several European Union countries in policies and actions on labor migration within the EU. Efforts have been made to ensure parties to the dialogue meet regularly to share experiences and discuss matters of mutual concern for the two regions in managing labor migration flows between their two regions (EC & IOM 2011a). In February 2011, it brought together representatives of several European countries, the European Commission, and the European Union External Action Service (EEAS) on the one hand, and Asian labor-sending countries of the Colombo Process on the other, for a meeting that resulted in some joint recommendations to improve the management of labor migration flows between the two sets of countries. It also committed participating states to holding a third meeting of the Asia-EU dialogue in 2012 (EC & IOM 2011b). Governments noted that, given the dynamic nature of migration between Asia and the EU, regular dialogue is needed to promote better understanding and collaboration on issues of labor migration.

---

[17] This African Institute for Remittance (AIR) project is under the leadership of the AU, in close association with the World Bank and the European Commission, also involving the African Development Bank and IOM. Retrieved from http://www.avrupa.info.tr/EEAS_News_Archive/Haziran2011,Eeas_20110603–1.html. Accessed 21 Nov 2011.

[18] See *Colombo Process: Asia-EU Dialogue on Labor Migration*. Available at: http://www.colomboprocess.org/index.php?option=com_content&view=article&id=63&Itemid=30. Accessed 21 Nov 2011.

(m) The EU-Latin American and Caribbean (EU-LAC)[19] dialogue is continuing to work on strengthening dialogue and cooperation between the European Union and Latin America and the Caribbean with a view to establishing management models on migration and development policies. The High Level working group meeting on migration in February 2011 resulted in a decision to implement a series of initiatives to improve national and regional capacity to produce and update data on migration. Through these initiatives, detailed knowledge of and data on intra- and inter-regional flows is to be collated and analyzed to inform the management of migration flows between the two regions. The project seeks to capitalize on the expertise acquired by migrants and enhance the institutional capacity in origin countries to better inform returning migrants about local employment and investment opportunities. This should result in more effective management of remittances for local community development initiatives.

Two questions arise from the above tour of inter-state and inter-regional efforts to enhance state cooperation on migration: To what extent are these disparate, largely informal and substantially variable actions laying the foundations for greater global governance on migration? Do some of these processes represent more effective models for labor mobility management, for example, than a more formal, global regime would?

As the above examples demonstrate, notwithstanding the informal and non-binding nature of these forums, countries in all regions are increasingly striving to improve capacity in migration management. Most of these initiatives do not relate to core migration and development concerns, such as remittances, community development, and diaspora engagement, but rather more specifically to migration. Nevertheless, these initiatives may invariably, even if only indirectly, influence policies and legislation on the well-being of migrants, the employment conditions of mobile workers, the development impacts of diasporas and the remittances migrants, and diaspora send home.[20]

This leads back to the concern of the GFMD Puerto Vallarta about how RCPs and IRF could best include the migration and development nexus, and the suggestion that a broader focus on capacity building could advance the discourse on this theme, and with time also influence actions in this regard. The issue of capacity is linked to ensuring that all stakeholders at the national level are engaged through inter-ministerial consultations. This ensures that the multifaceted nature of migration is taken into account in the management of migration nationally, and consequently how it is discussed in regional and inter-regional fora.

---

[19] See information on EU-LAC Summits at http://ec.europa.eu/external_relations/lac/index_en.htm. Accessed 21 Nov 2011.

[20] The Inter-governmental Consultations on Migration, Asylum and Refugees (IGC) were historically focused on asylum issues. Workshops convened by the IGC since 1985 were initially geared toward developing responses to the increased movement of asylum seekers in Western Europe. This focus has expanded over the years to include integration and the use of technology to improve migration management (Hallam and associates (2005, pp. 7–14).

Increasingly, governments appear to embrace the whole-of-government approach, irrespective of where they stand in the global migration landscape. Many are adopting multifaceted, comprehensive policies.[21] But they often do not have the capacity to put these into practice. Strengthening this capacity at national level is a prerequisite to the inclusion of migration and development issues on the agendas of RCPs and IRF. A concrete example of capacity building that goes in this direction is the establishment of the African Remittance Institute (ARI)[22] coupled with capacity building initiatives directed at Regional Economic Communities (RECs) in Africa, and initiatives to map the African diaspora address human trafficking and promote student mobility. All of these have been key initiatives of the Africa-European Union (EU) Partnership on Migration, Mobility and Employment.[23]

Similarly, actions at national level are influenced by regional and inter-regional discussions, as well as global fora such as the GFMD. The same national officials participate in all of these fora; hence, it is reasonable to expect mutual and reciprocal influences to occur between the national, regional, inter-regional, and global levels. With time, even if not set out as an objective for engaging in a RCP or IRF, policy coherence and harmonization may result, as has already been achieved in some cases (Hansen 2010, p. 26). The Philippines, as a member of several RCPs[24] and IRF as well as through its engagement in the GFMD, including as host of the 2008 GFMD, is arguably a good example of how engagement in multiple migration dialogue fora could over time shape a country's own policies. The Philippines is considered a leader in regulating recruiters, addressing irregular migration and providing welfare and protection to its emigrants. [25]

The limitations imposed by the non-binding nature of regional and inter-regional fora could be partially addressed by ensuring national institutions have the appropriate capacity to act, should they choose to, on the new knowledge and understanding gained from the dialogue processes they engage in. The informal nature of these fora ensures that their impact, even if positive, remains limited and will always be complementary, rather than determinative of national efforts (Hansen 2010, p. 43).

Following the second (GFMD 2008b, pp. 23–25) call for strengthened exchange and interaction between RCPs located in different regions, a global meeting of RCPs was organized in Bangkok in 2009. This drew together representatives from

---

[21] The approach taken by Portugal on immigrant integration is a case in point, where a whole-of-government approach helped develop an innovative and comprehensive action plan detailing 122 measures that outline the objectives and commitments of the various government agencies in welcoming and integrating migrants (IOM World Migration Report 2010, pp. 59–60).

[22] The ARI was launched in Addis Ababa, Ethiopia, in early June 2010.

[23] See EU and Africa in Partnership, available at: http://www.africa-eu-partnership.org/partnerships/migration-mobility-and-employment. Accessed 21 Nov 2011. The EU-Africa Partnership was born of the 2006 Joint Africa-EU Declaration on Migration and Development and provides an important platform for the incorporation of migration and development issues in the inter-regional agenda of the EU and AU.

[24] The Philippines is a member of the ADD, APC, Bali Process, and Colombo Process.

[25] IOM World Migration Report 2010, pp. 23, 38.

the principal RCPs around the world to share experiences and views and build informal networks and platforms for information exchange among them. For example, IOM hosts an Internet-based platform on its website to facilitate interaction between RCPs. A major obstacle to its effectiveness is the differential capacity of RCPs to provide updated information on their activities, which results in variable inputs. The Bangkok meeting also determined that global RCP gatherings needed to be held on a more regular basis. Consequently, a follow-up meeting of the principal RCPs and IRF was organized in October 2011, to focus on the role of capacity in enhancing cooperation on migration through dialogue (see footnote 8). Among others, this third global meeting of RCPs also sought to expose RCPs and IRF to emerging migration issues that may be of relevance to them, as well as stimulate reflection on the possible role that their respective groupings envision having with global fora for dialogue such as the GFMD and the HLD. The need for more frequent global RCP exchanges was again confirmed.

RCPs, IRF, and Global Cooperation on Migration

Clearly, the RCPs and IRF have contributed to progress in international dialogue and cooperation on migration over the past decade. At the national level, governments are improving their capacities to strengthen inter-ministerial consultations aimed at ensuring that all angles of what is truly a multifaceted issue are addressed. States' strong desire to maintain control over national policy (ICMC 2009, p. 4) has led to an intensification of bilateral migration governance frameworks, as evidenced by the mushrooming of bilateral labor agreements since the early 1990s. A few of these frameworks even factor development into the objectives of the migration agreement.

What is the likely way forward for global discourse and cooperation on migration and development? The GFMD is likely to remain the predominant global platform for states to deliberate regularly on migration and development issues. Modeled on the state-led, informal, and non-binding nature of the RCPs, the GFMD today plays a critical role in bringing together countries of origin, transit, and destination to deliberate on a wide range of migration and development issues. However, the GFMD is as yet insufficiently mature to have had any major concrete impacts. As Puerto Vallarta suggested, strengthening migration governance capacity at national level may be the key to ensuring more meaningful, more fully informed dialogue not only at regional level but also at the global level such as in the context of the GFMD.

In the absence of a binding, global regime for managing migration, RCPs and IRF have proven in a relatively short time to be effective facilities for this purpose. Despite their limitations, and the fact that they are informal and non-binding, they can and do contribute in appreciable ways to capacity building, policy coherence, and inter-state cooperation on migration. Strengthening the capacities of RCPs and IRF to interact more effectively, and deliver on their outcomes, may ultimately be a way of strengthening the GFMD and offering the elements for a more global form of migration governance.

Conclusion

As this chapter has illustrated, managing migration today depends on workable inter-state cooperation and partnership. Thus far, the RCPs and IRF have proven to be among the most effective vehicles for achieving these in practical ways among like-minded countries. The GFMD is emerging as the largest multilateral vehicle able to discuss such achievements in the broader context of migration and development.

Yet as currently structured, both remain insufficiently agile to fully meet the need for both international cooperation on migration and joint actions to leverage its benefits for development. There is increasing agreement that ever greater policy coherence in favor of a comprehensive and balanced approach to migration at national, regional, inter-regional, and global levels is necessary if the positive role that migration can play for origin and destination countries, and for the migrants themselves, is to be fully realized.

The innovations of the Swiss-led GFMD process in 2011 are laudable, and in line with changes proposed by several state and non-state actors over the years to ensure that GFMD deliberations progress from dialogue to action. The more decentralized approach of the global thematic meetings in 2011 has ensured greater focus on the possible practical applications on the ground of GFMD recommendations to date. It has also facilitated the participation of a broader range of government and non-state representatives at the various thematic meetings, a form of outreach not normally possible with the traditional GFMD summit meetings.

One of the notable outcomes of the GFMD V in 2011 was a broad acknowledgement within the GFMD that while regional processes on migration may bring their experiences to the GFMD, they remain distinct and independent of the GFMD. Governments actively engaged and committed to both the regional processes, and the GFMD by and large seemed to want to preserve the specificity and integrity of RCPs and IRF. Similar sentiments were expressed in earlier discussions in Puerto Vallarta that underlined the need for RCPs and IRF to remain distinct in terms of their local, "neighborhood" dynamics and shared strategic interests, much like regional trade agreements and economic integration processes. Nevertheless, the thematic meetings in 2011 generated considerable productive interchange between the GFMD and RCPs, and a number of the meetings involved members of the regional and inter-regional processes, with one resulting in a framework concept for regional cooperation intended for consideration by the relevant RCP and related IRF.[26]

It remains to be seen how the regional and the global processes – linked by common challenges, common structures, and common memberships but separated by different scales and agendas – can move forward in mutually complementary and reinforcing ways. The broader issue of the merit of these regional processes

[26] See the Report on the Workshop on Recruitment of Workers for Overseas Employment, Dubai, January 18–19, 2011 (www.gfmd.org).

nevertheless also remains: Are RCPs and IRF substituting for the lack of a global migration management regime? Laying the foundations for one? Or perhaps even delaying the coming into being of a global regime? Do these regional processes forestall true "globalization," for example, in terms of potential incompatibility between regional preferential behavior and global open markets? Or do they aid globalization? Whatever the responses to these questions, RCPs bear witness to the fact that managing migration today is often firstly an exercise in voluntary neighborhood cooperation.

RCPs may well have been derivative of the larger economic or trade integration processes, but they are yet to become the same force for international cooperation and action that the formal integration processes are today. For the most part, they remain formal enough to offer a frame for regional cooperation but not binding enough to command and enforce multilateral cooperation. They serve important practical purposes among neighbors but perhaps also help to delay any serious efforts at institutionalizing global cooperation in managing migration.

References

GFMD (2007a). Report of the First GFMD Meeting, Roundtable Session 3.4 on *Regional migration consultation processes and development: Advancing cooperation* (pp. 130–139). Retrieved from http://www.gfmd.org/en/documents-library/brussels-2007.html. Accessed 21 Nov 2011.

GFMD (2007b). *Regional consultative processes and development: Advancing cooperation* (Background Paper for Roundtable Session 3.4). Retrieved November 21, 2011, from http://www.gfmd.org/en/documents-library/brussels-2007.html

GFMD (2008a). *Regional consultative processes (RCPs): Inter-regional consultative fora and regional organizations and economic integration processes at the interface of migration and development* (Background Paper for Roundtable Session 3.3).

GFMD (2008b, October 27–30). *Protecting and empowering migrants for development* (Report of Proceedings for the GFMD Second Meeting). Retrieved November 21, 2011, from http://www.gfmd.org/en/gfmd-meetings/manila-2008.html.

GFMD (2009a, November 2–5). *Integrating migration policies into development strategies for the benefit of all* (Report of Proceedings for the GFMD Third Meeting). Retrieved November 21, 2011, from http://www.gfmd.org/en/documents-library/athens-2009.html

GFMD (2009b). *Regional and inter-regional processes and fora* (Background Paper for Roundtable Session 3.2).

GFMD (2009c). Report of the civil society days of the GFMD, Retrieved November 21, 2011, from http://www.gfmd.org/en/documents-library/athens-2009.html

GFMD (2010a). *How can Regional Consultative Processes on Migration (RCPs) and Inter-Regional Fora (IRF) best include the migration and development nexus?* (Background Paper for Roundtable Session 3.3). Retrieved from: http://www.gfmd.org/en/documents-library/mexico-2010.html. Accessed 21 Nov 2011.

GFMD (2010b, November 8–11). *Partnerships for migration and human development: Shared prosperity – shared responsibility*, 2010 (Report of Proceedings for the Fourth GFMD Meeting). Retrieved from: http://www.gfmd.org/en/documents-library/mexico-2010.html. Accessed 21 Nov 2011

GFMD (2010c). *Policy and institutional coherence to address the relationship between migration and development*. General Rapporteurs' Report to the Plenary (Ambassador Maria Bassols, Spain) on the Roundtable Session 3. Retrieved November 21, 2011, from http://www.gfmd.org/en/documents-library/mexico-2010.html

Hallam J. & Associates. (2005). *The inter-governmental consultations on migration, asylum and refugees (IGC): An assessment of 20 years*. Retrieved November 21, 2011, from http://www.iom.int/jahia/webdav/shared/shared/mainsite/microsites/rcps/IGC/2005-IGC-20-Years-Assessment.pdf

Hansen, R. (2010). *An assessment of principal regional consultative processes on migration* (IOM Migration Research Series, Vol. 38). Geneva: International Organization for Migration. Retrieved from http://www.iom.int/rcp. Accessed 21 Nov 2011.

ICMC. (2009). *Connecting the dots: A fresh look at managing international migration*. A report of the conversations on the global governance of migration. Geneva: International Catholic Migration Commission.

IOM. (2003) *Managing migration: Challenges and responses for people on the move* (World Migration Report). Geneva: International Organization for Migration.

IOM. (2010) *The future of migration: Building capacities for change* (World Migration Report). Geneva: International Organization for Migration.

Koehler, J. (2011). What government networks do in the field of migration: An analysis of selected regional consultative processes. In R. Hansen, J. Koehler, & J. Money (Eds.), *Migration, nation states and international cooperation* (pp. 100–127). Abingdon: Routledge.

Newland, K. (2010). How can Regional Consultative Processes and Inter-Regional Fora best include the migration and development nexus?, GFMD Civil Society Days Background Paper to Session 3.3, Global Forum on Migration and Development 2010, Puerto Vallarta.

Various Informal Reports of RCP Deliberations

Colombo Process. (2010). *Fourth ministerial consultations on overseas employment and labour migration: Migration with dignity*, 19–21 April 2011. Available at: http://www.colomboprocess.org/index.php?option=com_content&view=article&id=45&Itemid=26. Accessed 21 Nov 2011.

IGAD-RCP. (2010, October 20–22). *Migration management in the IGAD region* (Report of the Meeting). Retrieved from http://www.iom.int/jahia/webdav/shared/shared/mainsite/microsites/rcps/igad/. Accessed 21 Nov 2011.

MIDSA. (2010, November 15–17). *Managing migration through regional cooperation*. Report and Recommendations of the MIDSA Conference. Available at: http://www.migrationdialogue.org/midsa/. Accessed 21 Nov 2011.

Other Reports

Council of the European Union. (2005). *Presidency Conclusions – Brussels European Council*, 15/16 December 2005, 15914/1/05, REV 1. Accessed at 5914/1/05.

European Commission & IOM. (2011a, April 29–30). *Asia-EU dialogue on labour migration* (Report). Retrieved from http://www.colomboprocess.org/images/docs/asia-eu%20dialogue%20brussels%20proceedings%20final%20june%209%2008.pdf. Accessed 21 Nov 2011.

EC & IOM. (2011b, February 8–9). Asia-EU Dialogue on Labour Migration, 2011, available at: http://www.colomboprocess.org/index.php?option=com_content&view=article&id=63&Itemid=30. Accessed 21 Nov 2011.

ICMC. (2010, November 30). *The future of migration is now... a time to build governance*. Statement at the 99th IOM Council.

Chapter 14
Civil Society, the Common Space, and the GFMD

Chukwu-Emeka Chikezie*

Introduction

While hardly unique, the GFMD represents something of an experiment in international relations, including between the states and non-state players in the process. The vexed issue of what role the "civil society" should play in a process intended to be state-led, voluntary, and non-binding[1] was resolved at the outset in 2007 by creating a separate, but tenuously linked, process billed as the "Civil Society Day" (CSD, subsequently days as the initial 1 day became two).

Although the GFMD emerged in response to observations about a lack of institutional architecture for the global governance of migration,[2] it only partially fills this gap. The GFMD is, in essence, a forum for dialogue among governments. However, some have pointed out that it is neither totally inclusive in terms of cross-sectoral participation nor in terms of the migration topics within its purview (Betts 2011). Some have seen the Forum as typically more focused on economic issues,

* This chapter was prepared by Mr. Chukwu-Emeka Chikezie, Cofounder, African Foundation for Development (AFFORD, UK & Sierra Leone); Consultant with Up!-Africa Limited & Inclusive Growth Strategies Limited.

[1] As Matsas (2008) documents, the United Nations High-Level Dialogue on International Migration and Development generated no consensus on making international migration the subject of formal, norm-setting negotiations; thus, the GFMD – with its informal, state-led character, its limited role for the UN, international organizations, and civil society generally – was the compromise offered by the then UN Secretary-General to help build relationships of trust.

[2] "Global migration governance can be defined as the norms and organizational structures which regulate and shape how states respond to international migration" (Betts 2011). Compare also the definition used by Newland in the next chapter of this book.

C.-E. Chikezie (✉)
Consultant, UP!-AFRICA Limited, London, UK
e-mail: cechikezie@gmail.com

I. Omelaniuk (ed.), *Global Perspectives on Migration and Development: GFMD Puerto Vallarta and Beyond*, Global Migration Issues 1, DOI 10.1007/978-94-007-4110-2_14,

although since Athens (2009) the Forum debates have shifted more to a broader view of human development, including on rights and social protection.

The GFMD's informal, non-binding nature has helped avoid frictions among governments over the more sensitive aspects of the migration-development relationship but may not have satisfied those elements of "civil society" most focused on human rights, more specifically migrants' rights.

Each GFMD meeting has seen an evolution of the CSD generally in the direction of expanding the interaction between state and non-state actors. This culminated in what the Mexican GFMD 2010 Chair-in-Office billed as the "Common Space," an informal state–non-state debate on the podium. To what extent did the Common Space contribute to the GFMD process? Where to next for the role of non-state actors in the GFMD process and indeed for the "Common Space"? In attempting to answer these questions, this chapter makes four broad points:

1. The migration and development debate has given inadequate attention to the development dimension of the nexus, which appears to have been relegated to second place. Development actors with a stake in this debate have so far not seized the agenda, which may have implications for the potential usefulness of the process, perhaps even its efficacy in the longer term.
2. The term "civil society" has proven problematic as a moniker to embrace *all* non-state actors with a stake in the migration and development debate. Some seem to be more included than others; some are more powerful, better organized, and more vocal than others. The idea of the CSD being a space for all to stake their diverse and varied claims and agendas in the migration and development field still leaves much to be desired. Thus, the legitimacy of the CSD, and by extension the GFMD, may also be compromised somewhat by the civil society construct. Given the loaded and contested nature of the concept "civil society," part of the answer to enhanced inclusiveness and legitimacy may be to reframe the interaction and drop the term "civil society" in favor of one that carries less baggage ("non-state actors" may be more accurate but may also be less elegant).
3. There has arguably been more focus on the process of civil society involvement than on the outcomes. Judged on process, civil society activists typically give the GFMD a consistent "fail." However, when viewed through different lenses, the scale of civil society engagement in the GFMD process and participation in implementing recommendations from GFMD meetings becomes clearer. Mexico's Common Space innovation in 2010 was also primarily about process (perhaps understandably so, given the nature of many states' wariness about greater civil society involvement in the GFMD), but in achieving more inclusiveness, it was generally assessed to be a success. In thinking through the process of civil society engagement in the GFMD, this chapter argues for due consideration of the desired results, rather than to treat the process of civil society involvement as an end in itself.
4. Prior to 2011, the GFMD left little scope for an agenda-setting role by civil society or any other groups, who were generally expected to mirror or respond to the state agenda. However, GFMD 2011 set a different tone by leaving the civil society agenda-setting to civil society. It will be critical to draw the lessons from this in looking back at 2011.

Arguably, states could still lead the GFMD process but be more responsive to agendas set by non-state actors in a way that enhances the effectiveness and impact of the GFMD as a whole. Regardless of how well received the Mexican innovation of the Common Space was, or how well disposed subsequent GFMD Chairs may be to further enhancing this space, without addressing some of these wider deficiencies of state–non-state interaction in the GFMD context, the Common Space may well make little significant difference on its own.

Underpinning this argument and this chapter as a whole is the author's vision of migration in the twenty-first century as being about human mobility in all senses, and the natural flow and growth of human talent with an eye to the macroeconomic, sociopolitical, and cultural gains for the world, within a regulatory framework that is facilitative and enabling rather than restricting, controlling, and criminalizing. Clearly, both state and non-state partners are a long way from realizing this vision. The horrors of human trafficking, the risks of human smuggling, the unjust exploitation of human labor,[3] the realities of stigmatization and social exclusion of migrants, and the pain of family separation all point to a darker side of migration. Nonetheless, migration can still be the oldest poverty reduction strategy known to humankind,[4] and the role of governments and civil society should be jointly to build the sorts of institutions and processes that can continue and dignify this long-honored tradition.

It is an institutional and practical challenge for governments to devise transparent and participatory, but above all effective, mechanisms for the task at hand. And it challenges non-state actors to engage themselves in the process in ways that could effect real change.

The autobiographical box below seeks to share with the reader the experience of one diaspora actor's encounter with the GFMD. This experience of frustration and disappointment with the role of the civil society in the GFMD is shared more widely, far beyond migrant and diaspora groups and those with prominent ideological convictions.

Box 14.1 An Autobiographical Detour

I was born in the UK, grew up in Sierra Leone, and returned to the UK as a young man anxious and disillusioned about my job prospects in the middle of what subsequently came to be known as Africa's "lost decade" in 1984. Rather naively, I assumed that I would return to Sierra Leone at some stage, and that when I did I would somehow need to contribute to the country's development.

(continued)

[3] See, for instance, Hari (2009).

[4] See the observation by John Kenneth Galbraith: "Migration is the oldest action against poverty … it is good for the country to which they go; it helps to break the equilibrium of poverty in the country from which they come." Cited in Harris (2002, p. 119) and in *World Migration 2005* (International Organization for Migration 2005, p. 253).

Box 14.1 (continued)

It was these unspoken assumptions that led me, after a few years of working as an IT consultant, to pursue an academic degree that combined IT and development. As no such course existed at the time, I had to create my own. Some years later, in 1994, I co-founded the African Foundation for Development (AFFORD) in London, with a mission to expand and enhance the contribution of Africa's diaspora to African development.

Thus, I came to the migration debate through a development door. As a member of the African diaspora, I held quite an instrumentalist view of the diaspora's potential to contribute to development. Again, rather unconsciously, I have tended to think of development as the desired and purposeful outcome from the individual and collective application of human agency. In this light, Africa's development is fundamentally the responsibility of Africans themselves, both in Africa and the diaspora.

Through my involvement in AFFORD's work (as its Executive Director between 1999 and 2009), I found myself in New York in 2006 participating in a preparatory meeting for the UN High-Level Dialogue on International Migration and Development and later at the UNHLD itself. Separately to this, I had taken an interest in the concept of dialogue (being influenced by some of the writings of David Bohm).[5] I also saw the Secretary-General's speech[6] and report to the General Assembly on migration and development (United Nations General Assembly 2006) as a vindication of some of the work AFFORD had pioneered in championing the role of the African diaspora in Africa's development.

Expecting the meetings to at least explore in substantive detail how migration can contribute to development (and diasporas in particular), I was taken aback by the overwhelming thrust of civil society interventions that challenged the very premise of the HLD meeting itself. I can remember leaving the UN building and New York feeling both frustrated with and silenced by the broader civil society movement present. While the viewpoints of these civil society actors had legitimacy, their reluctance to countenance the idea that some migrants and diaspora groups actually welcomed this opportunity

(continued)

[5] David Bohm (1917–1992) was a distinguished American-born British physicist who is best known for his work on the fundamentals of quantum theory and relativity theory and their implications for other fields. He devoted his later years to the investigation of dialogue, which he described thus "Dialogue is 'not an exchange and its not a discussion. Discussion means batting it back and forth like a ping-pong game. That has some value, but in dialogue we try to go deeper…to create a situation where we suspend our opinions and judgments in order to be able to listen to each other.'" Retrieved July 24, 2011 from http://www.david-bohm.net/dialogue/experiment.html

[6] The Secretary-General Address to the High-Level Dialogue of the General Assembly on International Migration and Development, New York, September 14, 2006. Retrieved July 24, 2011 from http://www.un.org/migration/sg-speech.html

Box 14.1 (continued)

to explore how to enhance development prospects in their countries of origin was difficult to understand.

As the GFMD emerged from the HLD, this experience of frustration, and a sense of marginalization by the civil society movement itself, has accompanied every step of my involvement in GFMD civil society gatherings to date. Indeed, one is hardly alone in this. During the 2010 GFMD CSD in Puerto Vallarta, a group of migrants/diasporas who have been involved in development were moved to issue a statement to their civil society counterparts, asserting the important role diasporas play in development and calling for a more nuanced and balanced discussion.

This experience as a diaspora participant in the GFMD has provided an acute sense of where "civil society" falls short as a vehicle for the diverse expressions of the many stakeholders around the migration and development table. But concerns about the inclusiveness and fair share of voice in the GFMD CSD process extend well beyond any perceived marginalization of the diaspora. Some governments seem to have been keener to engage migrants and diasporas around the question of their contributions to development than have certain segments of civil society.

This chapter does not set out to prescribe a solution to the broader question about an ideal or optimal institutional architecture for the governance of global migration. It takes as a given the current GFMD arrangements and explores ways of improving its effectiveness in the context of the interaction between state and non-state actors. While the evolution of state–non-state interactions over the last 5 years of GFMD meetings forms a backdrop to its focus, this chapter makes no pretense of an in-depth or detailed analysis or assessment of those interactions or of civil society's engagement with the general migration and development debate. The chapter's purpose and purview are more narrow and pragmatic – what Mexico's innovative Common Space can tell us about the evolving state/non-state dynamic in the GFMD and its quest for more relevant, coherent policies to improve development prospects through migration, and migration outcomes through development.

The chapter interrogates the "migration and development" construct in terms of the relative absence and silence of development perspectives in the GFMD and examines the makeup of civil society participating in the GFMD meetings and proceedings and their involvement beyond the actual Forum meetings. It also considers the Mexican innovation of the "Common Space" in Puerto Vallarta in 2010. The concluding proposals are aimed at three groups: civil society actors who may feel excluded from the GFMD process but have something to gain and contribute through more proactive engagement; civil society actors already engaged but who may prefer narrow forms of participation when a broader engagement through multiple channels and stages might deliver more tangible results, and who, in espousing strong principles, may inadvertently be excluding others; and GFMD member governments with concerns about sharing such a space with civil society to discuss substantive issues. Governments could now consider more of an agenda-setting role for civil society in the GFMD process while preserving its essential state-led character.

Migration and Development?

Migration and development are complex and sensitive phenomena, which if placed together may often only increase the potential for confusion, contestation, and consternation. Matsas (2008) observes that, for some, a focus on development is merely the fig leaf that allows the international community to address in a multilateral setting the perennially touchy subject of international migration. De Haas (2008) places the current epoch of a focus on migration and development in an historical perspective. Pessimism and optimism about prospects for migrants to contribute to the development of their countries of origin have waxed and waned over recent decades.

According to de Haas, we are currently in an optimistic phase, coupled with a bout of collective amnesia about previous research into migration and development. De Haas argues convincingly that the migration and development nexus remains under-theorized, particularly on the development side. Thus, in the GFMD, there may well be a slightly breathless celebration of self-help and "bottom-up" development by non-state actors without adequate and nuanced consideration of the structural impediments to development or the critical role of the state.

Although we might expect de Haas, as a renowned scholar in this field, to make such criticisms and indeed to call for more empirically grounded studies to test hypotheses, development practitioners also share similar concerns about the absence of a robust migration and development "practice."[7] Seasoned GFMD officials also point to the fundamental lack of evidence on the concrete relationship between migration and development (Omelaniuk 2010).

However, while few would argue with the need for the migration and development debate to be academically grounded, theoretically informed, and empirically tested, and indeed for policy to be evidence based, this is also a political process. Thus, topics selected for inclusion in GFMD meetings reflect pragmatic choices informed by the degree of interest among participating states and the likely availability of some consensus leading to agreed outcomes and possibly even actions. Academic soundness may be only one consideration in this regard.

The initial strong interest in remittances in early GFMD meetings is a case in point. Activists generally criticized this apparent obsession with economic remittances (at the expense of a focus on migrants' rights) and for some the instrumentalization of migrants and their hard-earned resources.[8] Yet among states, there was a

[7] Private conversation with a senior US aid official, Washington DC, June 2011.

[8] See "PSI call for broad approach on Migration during GFMD." Retrieved June 3, 2011, from http://www.world-psi.org/TemplateEn.cfm?Section=Content_Management&CONTENTID=21496&TEMPLATE=/ContentManagement/ContentDisplay.cfm

clear policy consensus on the need to reduce the cost of remittance transfers by fostering greater competition, market transparency, and consumer awareness of their choices (Collyer 2011).

Arguably, policy makers may now be receptive to a renewed focus – within the development context – on jobs and ways that the migration and development nexus can help create more employment opportunities (Chikezie 2011). The ongoing global economic crisis and the elusiveness of strong jobs growth, even where economic recoveries are in evidence, may have created just such a policy opening. Gallup's first-ever world poll, published in 2007 and canvassing the views of citizens in over 130 countries, which was presented to CSD delegates at GFMD 2009 in Athens, found that "what the whole world wants is a good job." This emphasis on jobs is fully consistent with the conventional and state-endorsed understanding of development as reflected in the eight Millennium Development Goals (MDGs). Goal 1 is to eradicate extreme poverty and hunger and target 1B is to "achieve full and productive employment and decent work for all, including women and young people."

While there has been a measure of continuity in each successive GFMD meeting, there has also been change and expansion. For instance, the Mexican Government broadened the debate to focus on irregular migration, including its causes; migration, gender, and family; and the impact of climate change on migration and development (Government of Mexico 2010a). GFMD agendas have evolved since 2007 and are likely to continue to do so.

This bodes well for a more robust consideration of the development part of the nexus, but it is likely to happen only if a future GFMD Chair incorporates it as a theme for the year. Moreover, although GFMD Chairs enjoy some privileges in setting the agenda for the year, they need to secure the agreement and cooperation of members of the GFMD Steering Group (states only) and the Friends of the Forum (a wider group of states and some international organizations as observers).

There is something of the chicken-and-egg conundrum in all this. A weaker focus on development tends to result in a lower level of engagement from mainstream development actors and the marginalization of the subject generally. But a more rigorous focus on development (for instance, on the development impacts of migration policies and programs) may not happen unless there is engagement and support from mainstream development actors. Overcoming this conundrum is not just a challenge for the development actors.

Arguably, the strong emphasis on a subset of migration issues – border controls, security, and the return of unwanted migrants – reflects the priorities and briefs of the departments assigned to GFMD matters. Relatively few governments (such as France, Sweden, Switzerland, and the Netherlands) send development officials to GFMD meetings. This must clearly reflect the interests and priorities that these states attach to the GFMD. Therefore, if there is to be a rebalancing of the focus in favor of development (still within the context of the migration-development nexus), more states will have to initiate this through the anchor ministries and delegates.

GFMD and "Civil Society"

To address the question of how the Common Space innovation has or has not contributed to enhancing the efficacy of the GFMD process, we first need to examine what is meant by "civil society," and its involvement in the GFMD in the first place.

As already noted, the term "civil society" is highly contested and there is no single agreed definition. The World Bank uses the term "to refer to the wide array of non-governmental and not-for-profit organizations that have a presence in public life, expressing the interests and values of their members or others, based on ethical, cultural, political, scientific, religious or philanthropic considerations. Civil Society Organizations (CSOs) therefore refer to a wide array of organizations: community groups, non-governmental organizations (NGOs), labor unions, indigenous groups, charitable organizations, faith-based organizations, professional associations, and foundations."[9] Contested or not, this is a fairly conventional definition of civil society. GFMD member states have repeatedly recognized civil society as indispensable partners and mainstays in managing migration and development programs, supporting and protecting migrants and their families, and advocating for smarter policies. This recognition has extended to the private sector, but on the civil society part, the question of whether and how to involve the private sector in the CSDs has remained largely unresolved.

Few would dispute the centrality of the private sector to the migration and development debate – as employers, recruiters, facilitators, money transfer agencies, banks, finance institutions, airline companies, IT purveyors, etc. Indeed, civil society recommendations to the government delegates at GFMD 2010 in Puerto Vallarta, Mexico, included a call for greater integration of private sector actors and shared initiatives in migration and development (Government of Mexico 2010b). Yet over the years, some GFMD civil society activists have questioned whether the private sector has a legitimate place inside the civil society tent as a stakeholder (Rejšková 2011) or should rather be brought to the table only to be taken to task for the transgressions it is allegedly responsible for.

The private sector has generally shied away from engaging in the CSD and migration debates not least because of the political sensitivity of international migration in most countries of destination and because "civil society" has not been seen to be the right umbrella under which to engage.

One response by organizers of CSDs has been to involve business associations, such as Business for Social Responsibility (BSR) in Brussels in 2007 and in Puerto Vallarta in 2010, or the American Chamber of Commerce in Manila in 2008. Separate roundtable meetings with an agenda deemed to be more interesting to

[9] See World Bank. "Defining civil society." Retrieved July 16, 2011, from http://web.worldbank.org/WBSITE/EXTERNAL/TOPICS/CSO/0,,contentMDK:20101499~menuPK:244752~pagePK:220503~piPK:220476~theSitePK:228717,00.html

business were also organized in the run-up to the 2009 Athens GFMD.[10] One of the thematic meetings[11] in 2011 organized by the Swiss Chair-in-Office also succeeded in attracting a number of private sector companies to discuss labor mobility issues of significant concern to them. Another approach used in Mexico in 2011 was to affiliate with a prearranged business and migration-related meeting to address issues of relevance in the GFMD context. But even where an organization such as BSR plays an active role in facilitating business participation in the migration and development debate, it seems that a meeting of minds through dialogue still remains elusive.[12]

These side meetings have all been necessary precisely because the CSD framework has not sufficiently included all the non-state actors with a stake in the migration and development debate.

This chapter contends that just as there has been too narrow a definition of who or what constitutes "civil society" in the migration and development context, there has been too much of a narrow focus on participation in formal GFMD meetings. In reality, as Table 14.1 illustrates, there are several ways for non-state actors to be involved in GFMD work processes. Indeed, an assessment of GFMD achievements – roundtable GFMD proposals and outcomes as well as impacts on policy, legislation, and institutions – noted that by 2010, more than 50% of the 32 proposals for action from the Brussels and Manila GFMD meetings had been implemented by non-state observers and partners to the GFMD process (GFMD 2010). Clearly, states cannot act alone, and much of this implementation requires the active participation of non-state actors.

Part of the consternation may stem from the novel and perhaps nebulous nature of the GFMD. Using Newland's dichotomous characterization of international governance of migration as being either top-down or bottom-up (Newland 2005, p. 3), perhaps the GFMD is something of a hybrid, given that it involves states effectively engaged in a process of informal networking and dialogue. Not all of the points of entry and influence for non-state actors are immediately clear. And as Banulescu-Bogdan (2011) suggests, advocacy engagement with an entity such as the GFMD may demand of civil society actor's different skills and strategies from their usual ones. GFMD member states are more concerned with seeking practical solutions and effective policies than with norm setting, and this again presents challenges for conventional advocacy approaches.

Indeed, the objections that have typically been expressed by government officials to increased civil society engagement in the process seem to stem not from some

[10] See "The private sector and international migration policy." Retrieved April 4, 2011, from http://www.gfmd2009.org/UserFiles/file/MPI-IMEPO%20Athens%202009%20Business%20Roundtable%20SUMMARY.pdf

[11] Markets for Migration and Development (M4MD): Trade and Labour Mobility Linkages – Prospects for Development? September 13–15, 2011, Berne.

[12] For instance, see "How business can link migration and development," Retrieved June 12, 2011, from http://migrationlinkages.bsr.org/blog/how_business_can_link_migration_and_development

Table 14.1 Scope for non-state actors' involvement in GFMD processes

GFMD "activity"	Observations
Member states' pre-/post-GFMD consultations with domestic non-state actors	Whether these occur and who is invited to participate rather depend upon the democratic culture established and the government priorities in the country concerned, there may be limited scope for non-state national actors to shape the agenda
GFMD Chair-in-Office Themes Concept formation (with GFMD Steering Group and Friends of the Forum)	While there is no formal provision for non-state actors to engage in the process of agenda-setting for GFMD meetings, there is scope for informal exchanges to take place on the margins of meetings in Geneva. The fact that meetings take place almost exclusively in Geneva limits access to those civil society groups and international organizations with a presence in Geneva
GFMD roundtable background paper consultations and paper writing	Academic experts and international organizations are particularly well placed to play active roles here, although states are often keen to retain ownership over the process (which can be important for follow-through and implementation)
CSD planning committee meetings	An international advisory committee that seeks to be broadly representative of the diversity of non-state actors involved in the GFMD. CSD (including the private sector) assists the organization tasked with coordinating the CSD preparations
CSD roundtable background paper consultations and paper writing	Often, there is cross-fertilization of ideas (sometimes through use of the same people writing government and CSD papers) between the two sets of papers, facilitated in large measure in earlier years by the MacArthur Foundation sponsoring a roundtable of government and CSD paper writers in advance of the main GFMD meeting
CSD preparatory meetings	In some cases, these meetings have been used to engage especially the private sector in meetings deemed to be more relevant to them than the main CSD meetings
CSD	A wide cross section of mostly civil society actors attend these 2-day meetings; inevitably, budgetary constraints limit the number to about 200; the number of observers, including governments, has been on the increase in recent years
Government-CS "interface"/ Common Space	Disaffection with the "interface" that involved relatively few governments and/or civil society actors constrained by time and other factors gave way to the Mexican innovation of the Common Space involving semi-structured state–non-state interaction around topics of broader import to the migration and development debate than the specific GFMD themes
GFMD Government Days roundtables	Some experts and international organizations typically support governments in presenting evidence, serving as rapporteurs, panelists, expert witnesses, etc.; no formal access for civil society "groups" per se
Member state implementation of recommendations	GFMD recommendations are informal and non-binding, but member states rely heavily on a range of non-state actors for their effective implementation

(continued)

Table 14.1 (continued)

GFMD "activity"	Observations
GFMD ad hoc Working Group activities (studies, surveys, implementation, consultations, etc.)	As above
GFMD Platform for Partnerships (PfP) showcases and initiatives	Although governments are still expected to take the lead in highlighting projects of interest, there may be ample scope for non-state actors to work in partnership with governments on PfP initiatives[a]

[a]Upon the initiative of the Swiss Government, the PfP was established in 2010 for the principal purposes of ensuring some kind of memory "bank" of good practices linked to GFMD themes and discussions, showcasing GFMD follow-up projects, and encouraging cross-sectoral partnerships in any of these

Source: Author's observations

in-principle resistance to civil society per se but more from doubts that new or useful ideas will be forthcoming to move the process forward. After hearing civil society feedback and recommendations at the end of the 2-day CSD meeting in 2008 in Manila at the government-civil society interface meeting, the Philippines Chair, Foreign Affairs Undersecretary Esteban Conejos, thanked all the delegates profusely for their input but then lamented that he had actually hoped for something more useful to work with at the government meeting.

While the scope for non-state engagement and indeed influence within the GFMD process may exceed that implied by the general criticisms of the Forum by civil society activists, the monopoly on GFMD agenda setting that states retain does create something of a contradiction, given their subsequent dependence on non-state actors to implement the recommendations they adopt during their deliberations.

Common Space

Civil society exclusion from the main GFMD government meeting and the limited options for state–non-state interaction have been a contention for civil society activists since the inception of the GFMD process in 2007. Indeed, the 2006 UN meetings were a precursor to these tensions. As a result, different civil society groupings have adopted different strategies in response to the GFMD, characterized by one author as "insider-outsider" versus "outsider by choice" (Rother 2009). The compromises, tensions, and power plays that gave rise to the GFMD are well documented (for instance, see Matsas 2008; Roldan and Gasper 2010).

In broad terms, many migrant-sending developing countries had a preference for a UN-backed formalized, rule-based multilateral framework within which to address international migration issues. Many of these have already ratified the UN International Convention on the Rights of All Migrant Workers and Members of their Families. This is typically the position adopted by the more organized forms of civil society, such as labor unions, human rights organizations, and some migrant organizations. More powerful, migrant-receiving developed countries have resisted what they perceive to be encroachments on their national sovereignty and have favored a more flexible, informal dialogue process. These countries have typically not ratified the UN International Convention on the Rights of All Migrant Workers and Members of their Families (1990).

But the picture is not as black-and-white as these observations imply. Mexico has ratified the 1990 Convention on migrant workers, has strongly favored a UN-based multilateral system for addressing international migration (and development), but also hosted the GFMD in 2010. The complexities of human cross-border flows now mean that many countries – Mexico included – once perceived as primarily migrant-sending countries now have to redefine themselves also as countries of migrant destination and transit, facing the same challenges as the north. As one Mexican official pointed out in 2010, "we will not demand of the north what we cannot deliver to the south."

By the time Mexico took the GFMD reigns in 2010, the whole process of civil society involvement in the GFMD had undergone some incremental changes. In Brussels in 2007, the CSD was just 1 day, and a delegation of ten civil society representatives presented the CSD report at the opening session of the Government Meeting. Much of the debate generated at the time revolved around the identity and selection of those ten representatives, whom the organizers had preselected prior to the CSD (for security reasons, these names had to be submitted to the Belgian government for clearance well in advance of the meeting itself) from among the International Advisory Committee (IAC) that helped organize the CSD.

In Manila in 2008, the CSD had expanded to 2 days. Government delegates participated in a closing session of the CSD, and a handful of CSD delegates presented their conclusions to the opening session of the government meeting. Athens in 2009 followed a similar pattern, although the considerable distance between the location of the government and CSD meetings meant that few government delegates made it to the CSD interface session. Some 15 CSD representatives had around 20 minutes to present their findings at the opening session of the government meeting (without any feedback or discussion).

Seen in this light, what the Mexican Chair proposed and persuaded other governments to agree with was quite a radical departure from the norm.

In proposing to expand the time for government-civil society interaction, as well as the numbers involved (the only limitation was the capacity of the plenary hall in which the meeting took place), the Mexican government was careful not to encroach on the time traditionally allocated for government roundtable deliberations during GFMD meetings. The Mexicans sacrificed the morning "general debate" (that one official quipped was neither general nor a debate), typically dedicated to formal speeches, to accommodate the Common Space.

The term "Common Space" was chosen deliberately by the Mexican Chair to make a decisive break with the term "interface," which for some had little concrete meaning. In essence, the Common Space consisted of 2 hour-long sequential "Davos-style" six-person panel discussions with the panelists comprising government, civil society, and international organization representatives.[13] Each panel had a moderator who posed questions to the panelists and then fielded unscripted questions from the audience (Government of Mexico 2010b). Although the topics under discussion were substantial and topical – "Improving Public Perceptions of Migrants" and "Migration for human development – enhancing partnerships" – the Mexican government's objectives from the outset had been rather modestly limited to the political and symbolic rather than concrete outcomes related to these two themes.

Indeed, in brief remarks at the start of the Common Space session, Ambassador Juan José Gómez Camacho, who was most closely associated with shaping and driving the initiative, explained the goals of the Common Space:

1. To send a message from governments to civil society that they are ready to take steps to be more inclusive and transparent vis-à-vis the complexities of migration
2. For organized civil society to send a message to governments that it is ready to make a constructive contribution to discussions with governments (Government of Mexico 2010b)

Participants' feedback, as gleaned from feedback forms and Internet blog comments, was generally positive. One negative comment pointed to the fact that because there had been inadequate information about the Common Space in advance of the GFMD CSD, many civil society delegates left too soon to participate in the Common Space session. Nonetheless, as the room was filled to its 800-person capacity, this apparent oversight may have simply helped to reduce demand for places at the meeting to a more manageable level.

Some 116 partially or completed feedback forms were collected at the end of the session, and they suggest that most participants left satisfied. Table 14.2 below summarizes the responses.

Only two government delegate respondents seemed to disagree strongly on the usefulness of the Common Space event. One respondent felt a much more tightly focused discussion would have been preferable to what they felt were superficial and predictable comments that added no value to participants' knowledge. The other skeptical respondent similarly wanted more focus but also felt that the Common Space strayed too far from the GFMD's core mandate (without spelling out how or what the mandate was).

Many respondents lamented the lack of time for substantive discussion. A few wanted to hear more concrete proposals. Commentators clearly preferred the more free-flowing style of discourse to panelists reading from prepared scripts.

[13] A total of five government (including a Director-General of the European Commission), five civil society, and two international organization representatives participated in the two panels.

Table 14.2 Summary of responses from 116 Common Space evaluation forms

Response	Government (%)	Civil society (%)	International organization (%)	Other (%)
Panel one with score 3[a] or higher	84	94	89	86
Panel two with score 3 or higher	81	72	74	73
Agree, strongly agree, or very strongly agree that GFMD 2010 Common Space has helped raise my awareness about importance of fostering public debate about migration and development in my own country/context	84	78	68	82
Agree, strongly agree, or very strongly agree that GFMD 2010 Common Space has provided me with ideas for how I can generate public debate about migration and development in my own context[b]	82	72	63	82

[a]Respondents were asked to award 0 for not useful to 5 for very useful
[b]However, we have no data on post-GFMD Common Space follow-up at this stage

Box 14.2 Comments on the GFMD 2010 Common Space from Feedback Respondents

"Thank you for this very important opportunity. It was excellent and should continue. It shifted focus from formal speeches to a real exchange and touched on very critical issues" – civil society respondent.

"The Common Space should be conducted in a frank and courteous spirit and move towards concrete actions and results instead of being solely a sort of talk show" – civil society respondent.

"Common Space is a very good innovation which should become part of all GFMDs from now on. It provides space for informal discussions/viewpoints which is much more interesting and productive than formal, long presentations and statements. It is an excellent opportunity for government and civil society to discuss issues together" – international organization respondent.

"Need more time for substantive discussion for it to be meaningful" – international organization respondent.

"I think the model is excellent and served the purpose, it was very interactive and innovative. We should continue to engage both government and civil society in one forum – it was enlightening and illuminating, and most importantly, very frank and unedited discussions" – government respondent.

"Continue building the Common Space for improved dialogue to better address migrant challenges. Thanks a lot for this opportunity to be part of this global dialogue on this important development issue" – government respondent.

"As a debut for the GFMD the Common Space is a very good idea that should be carried forward by future chairs. Well done Mexico!" – unclassified respondent.

The Common Space succeeded in breaking the taboo in the minds of some that any prolonged encounter between governments and civil society would generate more heat than light. Measured against its specific goals, the Mexican government's innovation of the Common Space was a success. It showed that governments and civil society representatives can engage in constructive dialogue under the same roof within the GFMD framework.

But beyond this, the substance of the two topics set for discussion – perceptions of migration and partnerships – also mattered greatly for the Forum debates. The room for maneuver in migration policy that governments enjoy is usually circumscribed by domestic politics around migration, which in turn are shaped by several factors, among them public perceptions of migration. And progress in meeting any number of complex migration and development challenges, including changing public perceptions of migration, or galvanizing development agency support for migration issues, is inconceivable without robust partnerships. It is worth noting that these two topics mostly emerged from the Mexican Chair's own deliberations and informal soundings of a few non-state actors. Whether more relevant, dynamic debate could have been engendered through broader consultation on the Common Space themes is difficult to assess but may be worth considering for future Common Space encounters.

On the basis of the evidence available, there is certainly a case for subsequent GFMD Chairs to continue and further develop the Common Space idea. Indeed, the Swiss Chair sought to incorporate a Common Space into a number of its thematic meetings in 2011, as well as the Concluding Debate. The Common Space in the Concluding Debate, entitled "Looking at the Big Picture: Demographics, Youth (Un)employment, Development and Migration," certainly looks set to follow a similar purpose, pattern, and format (with modifications) to that established in 2010:

- Platform/panel shared by state and non-state actors.
- Panelists encouraged and at liberty to speak as individuals beyond the confines of their particular organizational or institutional mandates.
- Open, independent moderation aimed at constructive dialogue and solutions.
- Governments and a significant cross section of non-state delegates present in the audience.
- Question and answer interchange between audience and panelists, with the former strongly discouraged from making prepared statements, to engender free-flowing debate.
- Themes selected to move the debate beyond the more narrow and technical focus more typical of roundtable sessions are more "political" and demand more holistic, integrated approaches to problem-solving and creative leadership.

If Common Space 2010 managed to reduce some of the heat from the government-civil society interaction, the focus could now shift to generating more light. Arguably, both themes, "Improving Public Perceptions of Migrants" and "Migration for human development – enhancing partnerships" are worth continuing to explore in greater depth in the Common Space format. Finding new and creative responses to the challenge of public perceptions of migration – and not just in countries of destination, but across countries from where migrants originate, through which migrants transit, or to which migrants go, even though these may also be developing – is critical, unfinished business.

The issue of public perceptions of migration is something of a missing link in the panoply of migration policies and programs that is too little discussed (Canoy et al. 2006). Civil society organizations – with strong links not just to the grassroots of migrant communities but also more connected to the mainstream of host, origin, and transition societies (e.g., through faith community groups), diaspora groups, media organizations, academic institutions, think tanks, etc. – still have much to do to understand this phenomenon and to address it in constructive ways. The Common Space lends itself particularly well to this challenge.

Conclusion

Reflecting back on the 2010 Common Space experience at the GFMD meeting in Puerto Vallarta, Mexico, one cannot overlook the critical leadership role that the government of Mexico, as GFMD Chair-in-Office provided in securing all-round support for breaking with 4 years of "tradition." It was not a foregone conclusion that governments would agree to their proposal. The trust building with and cooperation of civil society leaders was also crucial. It required considerable political capital and sustained commitment from the senior ranks of the Mexican government, a willingness to take the necessary risks, and deft diplomatic skills to pull it off. It was a rare combination that came together to produce what may turn out to be a critical innovation to improve the overall relevance and effectiveness of the GFMD. The Swiss continuation of the experiment in 2011 signifies ongoing government and civil society support for the Common Space and may well mark its "arrival" as an enduring feature of the GFMD.

But a skeptic might ask, so what if there is now a Common Space within which governments, civil society, and international organizations can exchange views? On its own, does it amount to much? Are we not just creating a Tower of Babel? Without prejudging the outcome of the GFMD assessment process that will be discussed at the UN High-Level Dialogue in 2013, below are a few ideas that might spur a positive evolution of the Common Space.

First, development thinkers and practitioners need to contribute much more substance to the migration and development debate. The past few years have seen mounting interest in the roles that migrants and diasporas can play in development, but without a vision for development that envisages the necessary structural transformations for economies to offer real opportunities to growing ranks of unemployed and underemployed and some compelling ideas for how migration fits into this picture, it may be impossible to mobilize the required political support – in sending, transit, and receiving countries – for the tough decisions that are required. Also, without more concrete evidence of what works, and the development impact of programs and policies, the development factor in the migration and development equation will continue to be characterized more by hope, conjecture, and aspiration.

The GFMD has been characterized as a low-key forum that enables mostly technocrats to share ideas on best practice and workable solutions (Roldan and Gasper 2010). Indeed, some would attribute the success of the GFMD to date to this approach. But the story of migration is closely intertwined with the story of jobs – the departure in search of them, the jobs that are created through migrants' efforts in destination countries, the conditions under which migrants work, the prospects for work upon migrants' return, the jobs that can be created through migrants' and diasporas' transfers of knowledge, skills, investment, etc. And jobs are very firmly planted on the agenda of all countries, developed, developing, and transition. Thus, some more clear articulation of why migration and development matters today, linked to the issue of jobs and aimed at a skeptical public is vital.

There remains resolute resistance in some quarters to the very construct of migration and development founded, in part, upon the suspicion that this is a diversionary tactic plotted by increasingly parsimonious developed countries' treasuries seeking to shirk their aid obligations to the world's poorest nations and peoples. But across the developing world, millions of citizens are forging ahead, seeking solutions to their development challenges oblivious to these ideological debates that pass them by. Moreover, there is a very robust, ongoing debate about the limits to aid effectiveness (see, e.g., Easterly 2006) and indeed the real harm to developing regions' image when the focus is primarily on their aid dependence and vulnerabilities to shocks, such as famines, environmental disasters, etc. More attention to the human agency implied by migrants' and diasporas' own strategies helps to counter these impressions, without detracting from the responsibilities of states, both developed and developing.

For the space to be truly common, all non-state actors with a stake in the migration and development debate will need to "stake their claim." Diaspora groups no longer need to act like the clueless new kids on the block. Fresh from the parallel Peoples' Global Action on Migration, Development and Human Rights (PGA) in Mexico City where they felt their concerns and contributions as development actors were summarily dismissed with no more than a condescending ideological flourish, a group of migrants/diaspora active on the development front felt compelled to issue a statement in Puerto Vallarta at GFMD 2010, calling for issues above and beyond the fundamental one of migrant rights to be addressed by the CSD. "The emphasis on human rights alone risks obscuring migrants' sense of agency and the important experience and knowledge migrants contribute to policy makers and others. These contributions include, in part, development activities in their home countries."[14]

But it is unproductive to simply attend meetings and complain about being marginalized. Networks such as the recently formed Europe-wide Diaspora Platform

[14] Extract from the statement submitted at the CSD in Puerto Vallarta by migrant/diaspora actors.

for Development[15] could be proactive in approaching future GFMD Chairs with concrete and constructive proposals as to how they can contribute to the knowledge base, especially evidence of the impact of the contributions of migrants/diasporas to development.

Indeed, significant shifts within mainstream development thinking are ascribing a more central role to the private sector in the quest for development (e.g., see DFID 2011). Some are asserting that the private sector is not merely a stakeholder in development but a structural partner.[16]

The striking correlation between patterns of migration and trade suggest that both diasporas and the private sector should play much more central roles in the migration and development debates and the GFMD process as a whole. To be more specific, perhaps what is required at this stage is for a coalition of the willing comprising development practitioners (whether bilateral or multilateral agencies, NGOs, think tanks, etc.), academics, migrant, and diaspora groups primarily focused on enhancing development in their countries of origin, the private sector – employers of mobile talent, investors in developing countries, facilitators of transfers, etc., as well as policy makers and other actors focused on development to come together to contribute a sharper outcomes and impact-oriented dimension to the development dimension of the GFMD debate focused on 2012, 2013, and whatever emerges after the 2013 assessment process and UN High-Level Dialogue.

The key point is that a Global Forum on Migration and Development with a Common Space at its heart needs to give more space to development and be a true common space for all. While those with a stake need to demand their place round the table, it is also incumbent upon Chairs and organizers to ensure that this is sacrosanct. Those wishing to raise very legitimate concerns about the human, social, and environmental costs of globalization, neoliberalism, and growing inequality should do so but not to the exclusion of others with other pressing concerns, especially when those other concerns seek to tackle some of the very same concerns through different methods.

A second point relates to the GFMD itself, a rather complex entity with many moving parts, and the challenge that exists for more non-state actors to be more strategic in taking a step back to understand it in a more holistic sense with a view to influencing it.

Finally, while the GFMD meetings themselves have been the focus of the attention of civil society in particular, it is worth remembering that although the GFMD carries only light institutional baggage, beyond the meetings it involves some agenda setting by the incoming Chair which has to negotiate the focus and themes with other member states. There are also ongoing follow-up activities from previous meetings (carried out through the two ad hoc working groups).

---

[15] See "Formation of Europe-wide diaspora platform for development." Retrieved June 20, 2011 from http://www.migration4dayevelopment.org/sites/m4d.emakina-eu.net/files/press_release-communique_europe-wide_diaspora_platform.pdf

[16] Speech by Anne McCormick, Director, Corporate Affairs, Diageo Africa at Diageo plc to "Charting a future for business-donor partnerships for development" meeting, London, July 5, 2011.

Non-state actors necessarily play critical roles in the implementation of recommendations, best practices, etc. that are the focus of GFMD discussions. As one civil society representative speaking from the floor of the Common Space noted, "civil society acts as the canary in the mine," alerting government to dangers in the atmosphere. Without encroaching upon the sovereignty of states with the state-led GFMD process, it should be possible to take more structured soundings of all migration and development stakeholders in setting the agenda for GFMD meetings. The gain is likely to be more efficient and effective implementation and in the longer term greater impact of migration and development policies and programs.

As Betts (2011) argues, on a global level, there is a lack of vision and leadership in the area of global migration governance. Responses are ad hoc and there is a sense of drift. The critical leadership he calls for will necessarily come from a handful of states. But in our globalized, interconnected, interdependent world, such leadership is not for states alone. Mexico demonstrated commendable leadership in opening up the "common space," and Switzerland has carried that forward in 2011. It is time for more non-state actors to offer their own leadership to fill that space and make the GFMD more relevant and useful, with a long-term view to enabling the mobility of human talent and creating a more prosperous, more equal, more peaceful world.

References

Banulescu-Bogdan, N. (2011). *The role of civil society in EU Migration Policy: Perspectives on the European Union's engagement in its neighborhood*. Washington, DC: Migration Policy Institute. Retrieved July 27, 2011, from http://www.migrationpolicy.org/pubs/EUcivilsociety.pdf

Betts, A. (2011, March 29). *Policy primer: The UK and global migration governance*. The Migration Observatory at the University of Oxford. Retrieved July 24, 2011, from http://www.migrationobservatory.ox.ac.uk/policy-primers/uk-and-global-migration-governance

Canoy, M., Beutin, R., Horvath, A., Hubert, A., Lerais, F., Smith, P., & Sochacki, M. (2006). *Migration and public perception*. Brussels: Bureau of European Policy Advisers (BEPA), European Commission. Retrieved July 27, 2011, from http://ec.europa.eu/dgs/policy_advisers/publications/docs/bepa_migration_final_09_10_006_en.pdf

Chikezie, C. (2011, 29 June). Africa's development: the global bond. *Open Democracy*. Retrieved June 30, 2011, from http://www.opendemocracy.net/chukwu-emeka-chikezie/africa%E2%80%99s-development-global-bond

Collyer, M. (2011). *The development challenges and the European Union: EU-US Immigration Systems 2011/08*. San Domenico di Fiesole (FI): Robert Schuman Centre for Advanced Studies, European University Institute.

de Haas, H. (2008). *Migration and development: A theoretical perspective* (IMI Working Paper 9). University of Oxford: International Migration Institute. Retrieved April 14, 2011, from http://www.imi.ox.ac.uk/pdfs/imi-working-papers/WP9%20Migration%20and%20development%20theory%20HdH.pdf

DFID. (2011). *The engine of development: The private sector and prosperity for poor people. London: Department for International Development*. UK Aid: Department for International Development. Retrieved June 7, 2011, from http://www.dfid.gov.uk/Documents/publications1/Private-sector-approach-paper-May2011.pdf

Easterly, W. (2006). *The white man's burden: Why the west's efforts to aid the rest have done so much ill and so little good*. New York: The Penguin Press.

Global Forum for Migration and Development. (2010). *GFMD achievements to date*. Retrieved May 15, 2011, from http://www.gfmd.org/en/component/simpledownload/?task=download&fileid=ZG9jdW1lbnRzL21leGljby9nZm1kX21leGljbzEwX2FjaGlldmVtZW50c19wcHRfZm9mMl9yZXYwNG1heTEwLnBkZg%3D%3D

Government of Mexico. (2010a). *Partnerships for migration and human development: Shared prosperity – shared responsibility*. Fourth Meeting of the Global Forum on Migration and Development, Mexico 2010. Retrieved May 22, 2011, from http://gfmd.org/en/component/simpledownload/?task=download&fileid=ZG9jdW1lbnRzL21leGljby9nZm1kX21leGljbzEwX3JldmlzZWRfdGhlbWVfY29uY2VwdF9wYXBlcl9lbi5wZGY%3D

Government of Mexico. (2010b, November 8–11). *Partnerships for migration and human development: Shared prosperity – shared responsibility. Report of the Proceedings*. Fourth Meeting of the GFMD. Puerto Vallarta, Mexico. Retrieved May 28, 2011, from http://www.gfmd.org/en/component/simpledownload/?task=download&fileid=ZG9jdW1lbnRzL21leGljby9nZm1kX21leGljbzEwX3JlcG9ydF9vZl90aGVfcHJvY2VlZGluZ3NfZW4ucGRm

Hari, J. (2009, April 7). The dark side of Dubai. *The Independent*. Retrieved March 24, 2012, from http://www.independent.co.uk/opinion/commentators/johann-hari/the-dark-side-of-dubai-1664368.html

Harris, N. (2002). *Thinking the unthinkable: The immigration myth exposed*. London: I. B. Tauris.

International Organization for Migration. (2005). *World Migration 2005*. Geneva: IOM.

Matsas, R. (2008, June 5–7). *The global forum on migration and development: A new path for global governance?*. Paper presented at the 2008 ACUNS Annual Meeting: The United Nations and Global Development Architecture, Bonn, Germany. Retrieved May 31, 2011, from http://www.egmontinstitute.be/papers/08/sec-gov/080605-Matsas-Acuns.pdf

Newland, K. (2005). *The governance of international migration: Mechanisms, processes and institutions*. A paper prepared for the Policy Analysis and Research Programme of the Global Commission on International Migration. Geneva: Global Commission on International Migration. Retrieved July 26, 2011, from http://www.gcim.org/mm/File/TS%208b.pdf

Omelaniuk, I. (2010, March 29–30). *The GFMD and migration and development*. Paper presented at Population, Integration and Law: Implications for Immigration Policy. Retrieved March 31, 2011, from http://migration.ucdavis.edu/rs/files/2010/omelaniuk-the-gfmd-and-migration-and-development.pdf

Rejšková, T. (2011). *Paving the road for international discussions on migration – Global forum on migration and development: An interview with Stefan Rother*. Retrieved May 22, 2011, from http://arnold-bergstraesser.academia.edu/StefanRother/Papers/401021/Paving_the_road_for_international_discussions_on_migration_Global_Forum_on_Migration_and_Development._An_interview_with_Stefan_Rother

Roldan, B., & Gasper, D. (2010, March). The Global Forum on migration and development: "All talk and no action" or "a chance to frame the issues in a way that allows you to move forward together"? (Working Paper 497). The International Institute of Social Studies. Retrieved May 22, 2011, from http://biblio.iss.nl/opac/uploads/wp/wp497.pdf

Rother, S. (2009). "Inside-outside" or "outsiders by choice"? Civil society strategies towards the 2nd Global Forum on Migration and Development (GFMD) in Manila. *ASIEN 111* (95–107). Retrieved May 31, 2011, from www.asienkunde.de/content/zeitschrift_asien/archiv/pdf/111_rother.pdf

United Nations General Assembly. (2006).*Globalization and interdependence: International migration and development*. Report of the Secretary-General. Sixtieth session, agenda item 54 (c). Retrieved July 1, 2011, from http://www.queensu.ca/samp/migrationresources/reports/Report%20of%20the%20SG(June%2006)_English.pdf

Chapter 15
The GFMD and the Governance of International Migration

Kathleen Newland*

Introduction

States turn to international cooperation when unilateralism fails them, and they come to believe that their purposes can be better served by working with others. In this framework, global governance can be thought of as the set of collaborative practices that states and other actors adopt to bring about the best outcomes attainable and reduce the likelihood of negative outcomes materializing from global forces (Newland 2010). Global governance is "global" only in the sense that the forces whose effects it seeks to tame are global, not because it operates at the global level.

International migration is one of the forces that most governments have failed to manage satisfactorily through unilateral policies and programs, but international collaboration—especially between origin and destination countries—has also been weak, in large part because the debate on migration has been so polarized in international fora.

The Global Forum on Migration and Development (GFMD) broke through that particular stasis, in part by doing what previous framings of the issue failed to do: identify a common interest among states. The United Nations (UN) member states that participated in the General Assembly's High-Level Dialogue on Migration in 2006 were able to take the next step to continue discussion on migration in the GFMD because it was framed around an issue in which all countries professed an

* This chapter draws from earlier work on the governance of international migration that she has done for the UN Secretary-General's working group on Migration (2002), the Global Commission on International Migration (2005), and MPI's TransAtlantic Council on Migration (2011).

K. Newland (✉)
Cofounder of the Migration Policy Institute (MPI) and Director of its program on Migrants, Migration and Development, Washington, DC, USA
e-mail: knewland@migrationpolicy.org

I. Omelaniuk (ed.), *Global Perspectives on Migration and Development: GFMD Puerto Vallarta and Beyond*, Global Migration Issues 1, DOI 10.1007/978-94-007-4110-2_15,

interest—development—rather than on migration generally, with its many areas of contention among states.

Today, and for the foreseeable future, the actions that individual states can take on migration are not constrained by supranational institutions or by the "norms, rules, principles and decision-making procedures"—of international regimes but by *reality*. States—even those states accustomed to thinking of themselves as leading the international system—are buffeted by forces and prospects that defy unilateral control by even the most powerful.

These forces are global, whether they are private capital movements that can undermine a nation's economy or global warming that can destabilize the environment on which nations' productivity and quality of life depend. Another of these forces is international migration.

Evolution and Innovation in Global Governance of Migration over the Past 20 Years

To understand the place of the GFMD in the governance of international migration, it is useful to review the development of various frameworks for international debate and cooperation over the past two decades or so. The opening of the current chapter in the history of the governance of international migration can be (somewhat arbitrarily) traced to 1990, when the UN General Assembly passed a treaty on the rights of migrant workers and their families. It was an inauspicious start, since the treaty could not attract the 20 ratifications needed to bring it into force, and languished for 13 years before finally reaching that threshold (UNOHCHR 2011). Attention to international migration in the 1990s was sporadic and largely fruitless.

The substantive and pragmatic chapter on migration in the Program of Action formulated for the 1994 International Conference on Population and Development was endorsed by 160 governments and then more or less ignored in practice. Repeated calls for a UN conference dedicated to migration were rejected by the major migrant-receiving countries, which feared a bruising North-South confrontation over issues of access to their territories and labor markets and issues of migrants' rights. In 1996, a regional conference on migration in the former Soviet Union and contiguous countries was held in Moscow, but its impact was limited by basic, unresolved differences among the participating countries. The conference did, however, put in place some structures for continuing dialogue that have been useful: a steering committee of interested states continued to meet after the conference, up until July 2000. Subsequently, the conference process continued through four working groups on persons of concern, migration management, NGOs and civil society, and legislation. The CIS conference also left behind an invigorated and empowered NGO sector in the region, with greatly improved access to dialogue with governments (ICVA 2000).

No UN agency in the 1990s had migrants or migration processes as priorities—not even the International Labor Organization (ILO), with its historic mandate for the protection of migrant workers going back to 1919. Migration was seen as too political and divisive a topic for the agencies to tackle. The International Organization for Migration (IOM), which operates outside the UN system, was held to a fairly narrow portfolio of service provision by its member states. Despite the growing concern that migration was escaping the control of even the most capable governments, collaboration on common problems was confined to a handful of groupings of mostly wealthy countries, such as the Intergovernmental Consultations on Migration, Asylum and Refugees (IGC) whose membership consists of Australia, Belgium, Canada, Denmark, Finland, France, Germany, Greece, Ireland, the Netherlands, New Zealand, Norway, Spain, Sweden, Switzerland, the UK, and the USA (IGC 2011).

All of this changed quite abruptly around the turn of the millennium. Suddenly, migration was a common topic of international debate, in existing institutions and frameworks created for that purpose. Between 1999 and 2011, the UN High Commissioner for Refugees (UNHCR), the World Trade Organization (WTO), and ILO all put migration-related issues high on their policy agendas (through attention to "mixed flows," trade in services, and the International Labor Conference, respectively); several high-level consultations were established (including the Berne Initiative, the Global Commission on International Migration [GCIM], and IOM's International Dialogue on Migration); and the Migrant Workers Convention drew its twentieth ratification and came into force in 2003 (45 states are parties to the convention as of October 2011), triggering the creation of a new treaty-monitoring body serviced by the Office of the UN High Commissioner for Human Rights (OHCHR 2011).

In 2002, UN Secretary-General Kofi Annan established a working group in his office under Assistant Secretary-General Michael Doyle to examine and report back to him on options for the UN system to address international migration issues more systematically (United Nations 2002). Unable to propose a solution that it felt was politically feasible at that time, the working group recommended further study by a state-led deliberative body. The governments of Sweden and Switzerland took the lead in creating the GCIM, which met from 2003 to 2005. The commission recommended that an interagency group prepare the ground for an ill-defined "Global Migration Facility" (GCIM 2006).

The Bretton Woods Institutions (especially the World Bank) and several multilateral regional development banks discovered that migrant remittances surpassed the world's total official development assistance (ODA) and started a research effort to understand how these flows of funds could have a greater impact on development (Ratha 2003). Other regional development banks, in particular the Inter-American Development Bank, also devoted substantial attention to remittances.

Recognizing the regional character of most international migration flows, almost every region of the world, except for the Middle East and North Africa (MENA), established some kind of regional consultative process (RCP) on migration (Hansen 2010).

Despite this flurry of activity, it remained clear that the countries of immigration (which were also the major donors of development assistance and support for international institutions) had no interest in creating a new UN agency, or any other kind of global institution, to help govern international migration. Some argued that regional processes (where regional powers could set the agenda and dominate the proceedings) were the most appropriate setting for the discussion of international migration. Some also maintained that IOM, with its limited mandate to offer migration services, provided all that was needed to foster cooperation on migration issues. Many states of the global South continued to advocate in favor of bringing the discussion of international migration into the UN.

Confronted with this lack of consensus among states, UN Secretary-General Kofi Annan was forced to innovate, which he did using three mechanisms. First, Annan persuaded the General Assembly to convene a High-Level Dialogue (HLD) on Migration and Development, which took place in September 2006. Second, he established the Global Migration Group (GMG), which today comprises 16 international agencies working on some aspect of international migration, including IOM. Finally, he appointed a special representative, Peter Sutherland, to act on his behalf to move the process along. But the question remained: what kind of process was to be moved?

The story of the evolution of the governance of international migration over the last 20 years reveals two important lessons: one, that evolution of governance, like what we understand about the evolution of life on earth, moves in fits and starts. Long periods of stasis are broken by intense periods of activity, which may draw upon ideas long thought to be dead ends. A second lesson is that innovation is essential to overcome inertia, and often is spurred by one persistent individual in a key position—in this case, Secretary-General Kofi Annan, who drove the process of innovation for 5 years without seeing a concrete outcome. Nonetheless, he did not give up on the idea that the international system needed a way to deal with the last pillar of globalization that lacked any means of institutionalized cooperation among states.

The Creation of the GFMD

The outcome of the HLD in 2006 was a decision to establish the GFMD, outside the UN system. This outcome did not emerge spontaneously by the consensus of the majority of participating states. Rather, it was moved by a small group of enthusiasts, who included senior officials from a handful of member states, the Secretary-General and his staff, the Secretary-General's special representative for migration and his advisors, and a few discreet external migration specialists. They worked before and during the High-Level Dialogue to lay out a compelling rationale for a GFMD, design an operational mode for it, persuade other delegates to the HLD of its merits, craft compromises to overcome objections to the plan, and devise a way of introducing the concept as an outcome of the HLD that would be hard to turn down.

The scope of the GFMD was limited to the relationship between migration and development to gain the acquiescence of the major migrant-receiving states, which had no interest in rehashing the divisive debates on migrants' rights and illegal immigration that had polarized UN debates on international migration (although, as noted below, the strictures on the GFMD agenda were relaxed over time). The GFMD would be financed entirely by voluntary contributions. Refugee and asylum issues would not be taken up, in order to avoid overlap with the mandates of established institutions. The agenda would be broad enough, within the enumerated constraints, to accommodate the interests of many states.

Collaboration on developing various roundtable topics would also be voluntary, but would always involve states from both the North and the South. Conclusions and recommendations would be nonbinding. There would be no institutionalized secretariat that could come to control the agenda or products of the meetings. With these ground rules, states met for the first GFMD in 2007. By the second GFMD, in Manila in 2008, more than 160 states participated, and participation has remained high thereafter.

- The form of the GFMD has continued to evolve. After the first two meetings, two ad hoc working groups were established to help sift through the annual accumulation of recommendations and outcomes, and maintain the momentum of discussions between annual sessions of the GFMD; one deals with the protection and empowerment of migrants and the other with migration data, research, and policy coherence.
- To help promote, document, and showcase some of the outcomes of GFMD collaboration, and spur further international cooperation on good practices in migration policy, a "Platform for Partnerships" (PfP) was established in 2010. States could come together in the context of the PfP to collaborate on concrete projects. Three projects were launched initially: a project to create a handbook on diaspora engagement, a project on unaccompanied child migrants, and a data project to create "migration profiles" of selected countries. The choice of projects to be showcased and promoted on the PfP remains arbitrary, reliant largely on what individual governments are prepared to bring forward as their good practices. It does not reflect the full range and complexity of outcomes of the GFMD, although the number of projects presented is growing. If governments choose to use the PfP, it could become a vehicle for states to reach a critical mass of agreement on certain policy and program approaches.
- The lack of a stable support structure (or secretariat) revealed some disadvantages over time in terms of managing a complex budget from 1 year to the next, capturing and cataloguing the documentation and outcomes of the annual meetings, and maintaining the institutional memory to help states chairing the GFMD to organize it as a process rather than a string of annual meetings. Participating states thus made the decision in 2009 to establish a "light administrative support unit" with limited staff and mandate.
- The format of the GFMD changed between 2010 and 2011, with the Swiss chair shifting focus from a once-a-year single conference to a series of thematic workshops

in various world regions, capped by a concluding debate. The Mauritian Chair for 2012 opted to revert to the earlier approach in which governments and other partners formed teams to prepare roundtables on agreed themes for a final summit. The Mauritian Government did, however, retain some elements of the decentralized thematic meetings to draw value and contributions from other non-GFMD events around the world to enrich the 2012 Forum.

- An evaluation of the GFMD process was scheduled to take place in 2011, which may result in further innovations.

The establishment of the GFMD shows some of the traits that characterize innovations in governance of global issues, in particular the need for "champions" to develop new ideas and persuade other stakeholders to accept them. Another trait is the flexibility needed to make adjustments in the framework of a new mechanism as it is becoming established.

Governance is Not Just for Governments

What is most striking about the rising interest in the international governance of migration is that it is brought about by the actions of disparate actors who are not, in most cases, regarded as powerful, individually, in the international system. States find their migration realities shaped by the combined force of millions of migrants, who take their savings and often their lives in their hands to move; by employers who decide to hire migrant workers; by intermediaries who specialize in placing workers in jobs abroad; by smugglers who help those without authorization to enter other countries; by criminals who mislead and betray migrants into conditions of virtual slavery; by migrant communities that support the mobility of members as a form of insurance; by migrant families that decide to live together in a destination country even if they are not officially allowed entry; and so forth.

The GFMD is distinctive among international fora in that, even though it is "owned" and led by states, it has structured a place in its deliberations for some of these nonstate actors through the participation of representatives of "civil society" (see Chap. 14).

States and nongovernmental institutions too often take a reflexively adversarial stance toward each other on migration issues. The governance of international migration would benefit from greater communication and cooperation between them, while still recognizing their differences. Migrant organizations are an unparalleled source of information and ideas about how migration processes actually work. Private businesses understand the day-to-day workings of the labor market in a way that governments do not, and public institutions have a longer-term and broader view of labor market dynamics than many businesses.

Many governments and international organizations do have institutionalized means of consultation with nongovernmental bodies, but some of these have become empty rituals. Public institutions need to reach out constantly to new partners and insist that the time invested on both sides has real value—beyond checking the box for civil-society engagement. For both practical and normative reasons, the views and knowledge resources of civil-society organizations, private businesses, migrant organizations, and expert bodies should be incorporated into policy discussions on international migration. For this to be productive, however, nongovernmental institutions need to elevate their experience and networks to a rigorous evidence base for policy formulation.

Knowing what works, what does not, and what motivates migrants and their communities of origin is an essential element of governance. Nongovernmental institutions need to be transparent about their representativeness, the sources of their expertise, and their goals. As monitors, evidence gatherers, agenda setters, and problem solvers, they can be serious partners of public institutions without giving up either their own priorities or their idealism.

The place of nonstate actors in the GFMD has evolved over the course of the first 5 years, but from the very beginning, the Forum included a Civil Society Day (2007) or days (subsequently), held immediately before the intergovernmental meeting. In the first two GFMDs, in 2007 and 2008, the civil society meeting merely reported, very briefly, to the intergovernmental meeting. In Athens in 2009, this perfunctory reporting was replaced by an "interface" between government and civil society, which allowed for debate between the two on the topics discussed at roundtables at both the civil society and governmental meetings. In some cases, for example, in the discussion of Roundtable 1.2 on "engaging diasporas and migrants in development policies and programs," the governmental participants in the interface adopted the recommendations from the civil society roundtable and presented them as governmental recommendations as well. In 2010 in Puerto Vallarta, the Mexican Chair in office organized a "Common Space" that brought civil society and governmental representatives together on an equal footing in two "Davos-style" panel discussions that allowed for a consistent experience for all the GFMD participants, although without specific outcomes. A common space is also planned for the 2011 GFMD.

The increasing interchange between the civil society and governmental participants at the GFMD illustrates what should be a broader principle in the governance of international migrations: that governance is too big for governments alone. The involvement of civil society has given rise to considerable soul searching in the GFMD (see Chap. 14). The relatively low participation of the private sector, however, in international discussions of migration governance or migration and development has been even more troubling. The Swiss GFMD in 2011 tried to correct this by calling together interested governments and business leaders from the World Economic Forum (WEF) network, international manpower agencies, social partners, and other relevant institutions in Berne in September 2011 to examine the migration, trade, and development nexus and develop innovative joint approaches.[1] Nevertheless, business participation remained disappointing.

[1] GFMD 2011 thematic meeting on "Markets for Migration and Development (M4MD): Trade and Labour Mobility Linkages – Prospects for Development?" Bern, September 13–15, 2011.

Moving the Debate on Perennial and Merging Topics in International Migration

By 2011, several issues (some longstanding and some new) had emerged in international policy discussions, including but not confined to the GFMD, as the ones most in need of international governance:

- The need to facilitate mobility (in particular, circular migration) for transnational citizens (those who routinely divide their work, family life, residence, etc., between two or more countries)
- The dearth of legal migration channels for people with low- or midlevel skills, despite persistent demands for their labor
- The lack of reliable information about the movement of people across borders and the composition of immigrant communities
- Methods of preventing the exploitation and endangerment of migrants
- Social and financial protections for workers involved in temporary labor migration
- The separation of families as a norm of migration management
- The potential of diaspora populations to contribute to the development of their countries of heritage
- The lack of widely observed minimal standards of employment and protections for low-waged migrant workers, especially domestic workers
- Persistent "brain waste" among migrants and the associated need to streamline credentialing for skilled immigrants
- The possibility of large-scale "survival migration" associated with dysfunctional economic policies, poor governance, armed conflict, political repression, high population growth in very poor countries, resource scarcity (in part the effect of global warming), and very unequal distribution of income in several regions
- The erosion of the asylum system established after World War II
- The penetration of organized crime into the migration "industry" and the integration of human trafficking into its broader "business plan"
- The need for new modes of protection-oriented collaboration on mixed flows (labor migrants traveling with refugees), stranded migrants, trapped migrants, third-country nationals, and other groups of migrants that do not fall into standard categories

The GFMD has addressed most, although not yet all, of these high-priority issues. Several topics of interest to governments have gained momentum through the GFMD, by being repeatedly addressed in roundtables. These include labor mobility, circular migration, temporary migration, diaspora engagement in the development of their countries of heritage, and the gaps in migration data. Not all participating states are interested in all of these topics, but one of the advantages of the GFMD is that its loose roundtable structure permits states to cluster around certain topics voluntarily, without creating pressure for others to engage.

By revisiting certain topics and creating voluntary communities of interest (which also include nonstate actors), concepts appear to have been clarified—although circular and temporary migration are still commonly conflated—examples

Table 15.1 Topics appearing on the agenda of more than one session of the GFMD, by year

Topic	2007 Brussels	2008 Manila	2009 Athens	2010 Puerto Vallarta	2011 Geneva
Temporary migration	*	*			
Circular migration	*	*	*		
Irregular migration		*		*	*
Role of nonstate actors	*				*
Diaspora engagement	*	*	*		
Gender and family			*	*	*
Reducing costs of migration				*	*
Protecting migrant rights		*	*	*	
Mainstreaming migration in dev.			*		*
Measuring and assessing impacts	*			*	*
Strengthening data		*	*		*
RCPs and IRFs		*	*	*	

Source: Assembled by the author from GFMD agendas 2007–2011
Note: Some of these topics were discussed in sessions that did not explicitly mention them in the agenda, so may be under-represented by this method of selection

of successful policies and programs are being shared, and joint actions agreed, although not always explicitly in the context of the GFMD. Table 15.1 shows the issues that have been on the agenda of the GFMD for more than 1 year.

Cause and effect are difficult to trace in policy debates, but action on a few of its chosen issues has coincided with repeated debate in the GFMD, often based on policy papers produced or commissioned by the governments organizing a roundtable discussion on the chosen topic. Innovative action is also supported by new policy-related research developed by participating governments and international organizations (which have observer status at the GFMD) including, for example, the compilation of handbooks of good practices on temporary labor migration, integrating migration into development practice, and diaspora engagement; and studies on the effectiveness of Migrant Resource Centers, social protection for temporary labor migrants, low-cost loans for migrants, methods for lowering the costs of recruitment, and the impacts of temporary labor mobility on families back home, among others.

The OECD, IOM, and ILO prepared a Compendium of Good Practice Policy Elements in Bilateral Temporary Labour Arrangements as a GFMD reference text, at the request of the Spanish chair of a roundtable session on this theme in 2007. More recently, in the context of a GFMD roundtable in Mexico, the governments of the Netherlands and United Arab Emirates commissioned studies on low-cost loans for migrants, regulating the migrant recruitment industry, the utility of Migrant Resource Centers, and social protection options for temporary labor migrants (GFMD 2010). IOM is helping to implement both the "Handbook on Engaging Diaspora in Development Activities" (with the Migration Policy Institute) and the "Migration Profiles" projects that have been placed on the GFMD's Platform for

Partnerships (PfP), while UNICEF is collaborating on a third PfP project on unaccompanied migrant children.

Among the actions that have been, arguably, influenced by GFMD-related research and debate are pilot projects on circular migration, new forums and institutions to draw diaspora groups more closely into development activities, new legislation and programs to protect migrants in domestic work, and initiatives to reduce the cost of migrating.

The fact that some of the more contentious migration issues, such as irregular migration and family unity, have been discussed at the GFMD, particularly at the fourth forum in Puerto Vallarta (2010), demonstrates that the GFMD has helped to bring about greater trust among governments and a resulting willingness to discuss difficult issues in a nonconfrontational manner.

The issues outlined above and others will play out against a backdrop of broader challenges that will confront states over the next decade, and quite possibly beyond. A slow and erratic recovery from the global financial crisis of 2008 is likely to keep domestic unemployment high and prolong the fiscal crises of many major countries of destination. This is likely to exacerbate the backlash against immigration and immigrants, which has fueled the rise of right-wing parties across Europe, generated floods of anti-immigrant legislation in the USA, and manifested itself in anti-immigrant violence in both North and South. It will also mask part of the underlying dynamic of global migration for the coming century and therefore delay policy responses to the aging populations and shrinking labor forces of a large number of wealthy industrialized countries. Climate change may reduce productivity and resilience in many parts of the world, making a significant number of states unable to absorb many new migrants. Terrorism is likely to persist, absorbing the attention and resources of governments.

International cooperation on migration will be increasingly important in the face of these broad challenges, and fora such as the GFMD where states can come together to build collaborative actions are likely to show their value more and more clearly.

Building Policy Networks

Even where the need for greater cooperation and coordination among states is acknowledged, a supranational organization is not seen by many states as the preferred avenue for the governance of international migration. The small number of states that have been willing to ratify international legal instruments on migration speaks to the lack of consensus on a normative level.

As an alternative, *international governance from the bottom up* stitches together the common threads of governmental responsibilities for problem-solving purposes. It is built on the basis of intensive interactions among government officials (bureaucrats, regulators, legislators, judges, and so forth), with similar functional portfolios—often interacting with nongovernmental stakeholders such as employers, human-rights advocates, and experts. These "policy networks" have been described by Professor Anne-Marie Slaughter (2004, p. 16) as the building blocks of world order for the age of globalization.

Formal, set, legally binding, and negotiated agreements are not the modus operandi of government networks. Rather, government officials continuously devise (and may revise) terms of agreement and are themselves responsible for implementing them. Coordinated solutions to common problems are sought through the constant exchange of information, the development of best practices, and the formulation of nonbinding codes of conduct—these are, of course, the preferred working methods of the GFMD.

Some of the most valuable products of a consultation process are, thus, intangible. Building trust and developing habits of cooperation among governments may be the GFMD's most valuable contribution to governance in the medium-to-long term. Nonetheless, setting benchmarks, even though nonbinding, would be useful to sustain the momentum of cooperation among GFMD participants. A plan of action (or some equivalent) on a particular topic would allow states to keep track of progress toward their agreed goals. As the policy networks created or reinforced by the GFMD grow, the participating states may be able to establish more concrete goals. In doing so, unanimity is not required. Moisés Naim (2009) coined the term "minilateralism" for a new kind of coalition of the willing formed around particular issues. He describes it as:

> … a smarter, more targeted approach: We should bring to the table the smallest possible number of countries needed to have the largest possible impact on solving a particular problem. Think of this as minilateralism's magic number. The magic number, of course, will vary greatly depending on the problem. Take trade, for example. The Group of Twenty (G-20), which includes both rich and poor countries from six continents, accounts for 85% of the world's economy. The members of the G-20 could reach a major trade deal among themselves and make it of even greater significance by allowing any other country to join if it wishes to do so. Presumably, many would. Same with climate change. There, too, the magic number is about 20: The world's 20 top polluters account for 75% of the planet's greenhouse gas emissions.

What is the magic number for international migration? A multilateral approach to mobility would need to engage the major countries of destination and origin to identify real problems that are widely recognized by these crucial players. The GFMD permits this kind of voluntary cooperation among groups of states. If they are successful in making progress toward their own goals through cooperation, they are likely to attract new collaborators. Strong policy networks created through a framework such as the GFMD make it possible for potential partners in cooperation to identify each other.

Conclusions

States have not yet reached a consensus on the overarching purpose of international cooperation on migration, but they have come to recognize some common problems and needs. The search for enhanced multilateral governance of international migration should begin with a consideration of what functions it would need to carry out for the good of countries of origin, destination, and transit, and for migrants. A platform for the development of international governance of migration will have to begin with a limited number of widely agreed functions that respond to the felt needs of states and address real issues in the societies affected by migration. A short list to begin with would likely include the following nine functions:

- Data collection, dissemination, and analysis; monitoring of trends
- Policy research and development
- Technical assistance and training (capacity building)
- A platform for discussion
- Promotion of migration-related development initiatives
- Support for negotiations that members choose to enter
- Antitrafficking initiatives
- Coordination of cooperative programs at the international or regional level
- Provision of certain financial, capacity-building, and logistical services to governments and migrants

The GFMD is already involved in the first five of these functions, but not in the last four. This may demonstrate that the participating states have decided to use it for some of the functions of governance (keeping in mind the definition of governance used for the purposes of this discussion) and may add more, or it may indicate that they have only limited interest in a more comprehensive platform for collaboration, or indeed that they do not see the GFMD as the preferred platform. But at this stage of the evolution of governance of international migration, the GFMD does not have many rivals at the global level.

To move toward more systematic frameworks for cooperation on international migration, national governments and other actors will have to acknowledge that different states have different goals, to compromise where possible, and to start building on recognized, common objectives that lead to mutual benefit from migration. These common objectives include the need for more knowledge and a better understanding of migration phenomena, fewer deaths of migrants in transit, a reduction in the influence of criminal networks, the minimization of tensions between migrant and host communities, greater safety and dignity for migrants, increased national security, easier mobility for authorized migrants, and stronger state capacity to implement policies that have been adopted.

The long-term success of the GFMD as an instrument of international cooperation will depend in part on the ability of its successive chairs and other governing structures to identify the issues of greatest salience to states that are *also* "ripe" for collaboration. This is the same logic that allowed the GFMD to be created, and implies a recognition of limits to what international collaboration will bear at a given time.

Similar forbearance is likely to be required from the participating states on the institutional framework of the GFMD as it evolves. The relationship of the GFMD to the UN may evolve, but if the GFMD is forced into the UN system before a strong consensus supports the move, the GFMD may lose its utility. UN agencies and other multilateral organizations do, however, help the GFMD participants to implement its decisions. As described above, many of the studies, compendia, and handbooks resulting from successive GFMD meetings have been completed by GMG member organizations such as IOM, ILO, and the World Bank, as well as by academic experts.

The expression of states' collective interest in particular migration issues through the GFMD may encourage international organizations to develop common agendas and activities on these issues. In 2010, for example, the Global Migration Group collectively produced a Handbook on "Mainstreaming Migration into Development Planning," which is being piloted in a number of countries by UNDP in partnership with other GMG organizations. Similarly, the collective interests of governments expressed via the GFMD may provide to researchers and analysts who hope to do policy-relevant work some indicators about the topics on which governments are seeking input. More importantly, the GFMD may continue to open more inclusive channels of communication between governments and other critical stakeholders in migration—in civil society, the private sector, migrant, and diaspora communities—without whose cooperation international migration cannot be truly governable.

In its fifth year, it is undoubtedly too soon to say whether the GFMD may be leading toward more robust governance of international migration, or whether states will choose it as the framework for developing their collaboration in this field, and other stakeholders will develop confidence in it. States may prefer to develop regional frameworks, for example, rather than a global approach, and deal with migration flows outside a given region through interregional discussions. Alternatively, states may decide, even if by default, to live with the relative anarchy of international migration as it proceeds today. But the argument for a set of collaborative practices that states and other actors adopt in order to bring about the best outcomes attainable and reduce the likelihood of negative outcomes is compelling and the stakes for states and for people who migrate is very high. The GFMD is at the very least a promising instrument for the further development of governance in the field of international migration.

References

GFMD. (2010, May 4). *GFMD achievements to date*. Retrieved from http://www.gfmd.org/en/recommendations-follow-up-actions.html

Global Commission on International Migration (GCIM). (2006). *Migration in an interconnected world: New directions for action.* Geneva: GCIM. Retrieved from http://www.gcim.org/en/finalreport.html

Hansen, R. (2010). *An assessment of principal regional consultative processes on migration* (IOM Migration Research Series 38). Retrieved from http://www.iom.int/rcp

ICVA. (2000). CIS conference process to continue. *Talk Back: The Newsletter of the International Council of Voluntary Agencies*, *2*(4). Retrieved from http://www.icva.ch/doc00000165.html

IGC. (2011). *Intergovernmental consultations on migration, asylum and refugees.* Retrieved October 24, 2011, from http://www.igc.ch/

Naim, M. (2009, July/August). Minilateralism: The magic number to get real international action. *Foreign Policy, 173*, 136.

Newland, K. (2010). The governance of international migration: Mechanisms, processes, and institutions. *Global Governance, 16*(2010), 331–343.

Ratha, D. (2003). Workers' remittances: an important and stable source of external development finance. In *Global development finance 2003* (chap. 7). Washington, DC: World Bank.

Slaughter, A. M. (2004). *A new world order*. Princeton: Princeton University Press.

United Nations. (2002, December 20). *Office of the Secretary-General unpublished report to the Secretary-General on migration* ("Doyle Report").

United Nations Office of the High Commissioner for Human Rights. (2011). *International convention on the protection of the rights of all migrant workers and members of their families.* General Assembly resolution 45/158 of 18 December 1990. United Nations Treaty Collection Databases, Retrieved October 13, 2011, from http://treaties.un.org/Pages/ViewDetails.aspx?src=TREATY&mtdsg_no=IV-13&chapter=4&lang=en

Index

I. Omelaniuk (ed.), *Global Perspectives on Migration and Development: GFMD Puerto Vallarta and Beyond*, Global Migration Issues 1, DOI 10.1007/978-94-007-4110-2,

MIX
Papier aus verantwortungsvollen Quellen
Paper from responsible sources
FSC® C105338

If you have any concerns about our products,
you can contact us on
ProductSafety@springernature.com

In case Publisher is established outside the EU,
the EU authorized representative is:
Springer Nature Customer Service Center GmbH
Europaplatz 3, 69115 Heidelberg, Germany

Printed by Libri Plureos GmbH
in Hamburg, Germany